CW01677204

THE RACE TO THE CAPE

UNION-CASTLE LINE
1857 - 1977

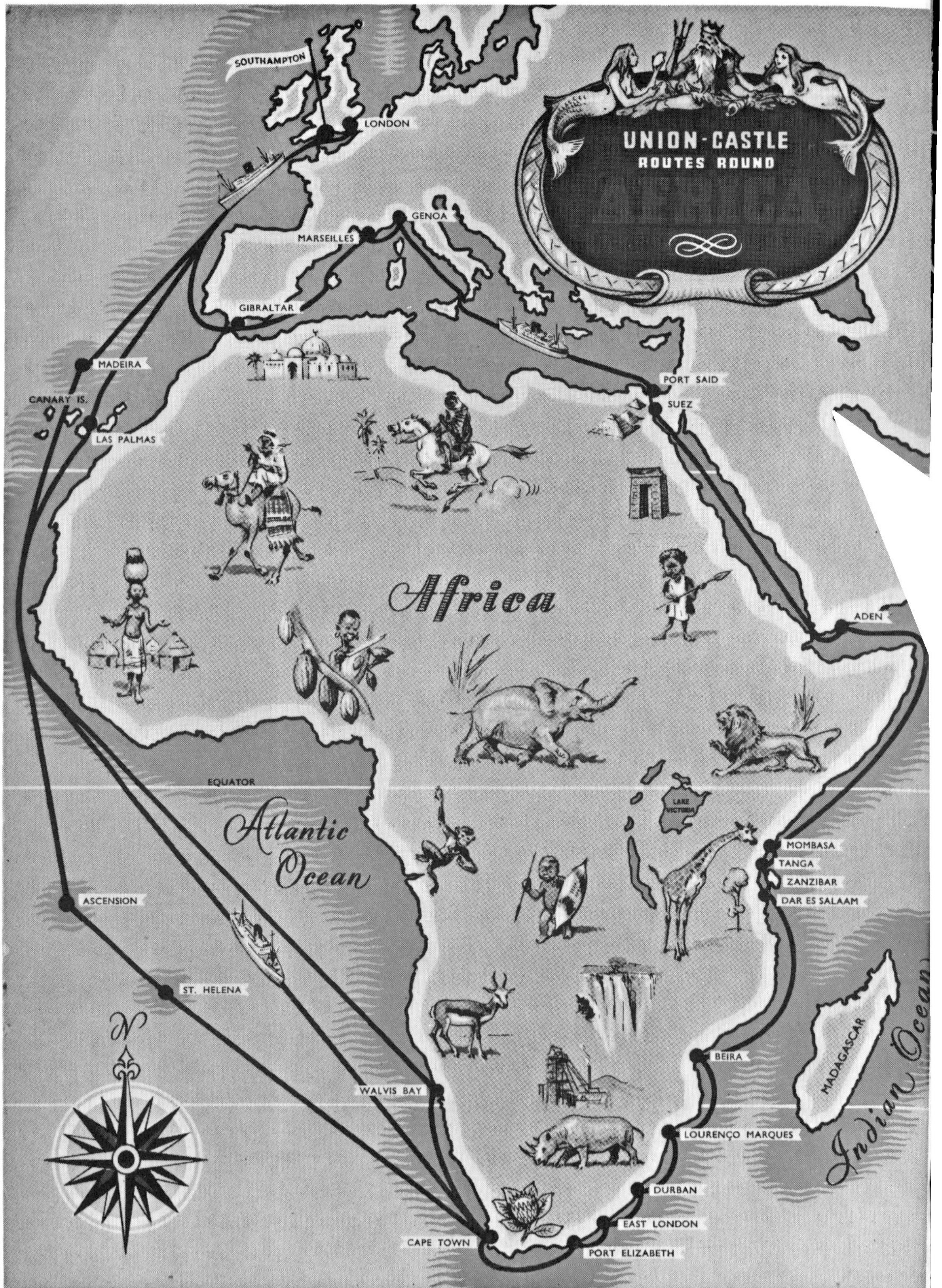
UNION-CASTLE
ROUTES ROUND
AFRICA
SOUTHAMPTON
LONDON
GENOA
MARSEILLES
GIBRALTAR
MADEIRA
CANARY IS.
LAS PALMAS
PORT SAID
SUEZ
Africa
ADEN
EQUATOR
Atlantic Ocean
MOMBASA
TANGA
ZANZIBAR
DAR ES SALAAM
ASCENSION
ST. HELENA
MADAGASCAR
Indian Ocean
BEIRA
WALVIS BAY
LOURENÇO MARQUES
DURBAN
EAST LONDON
CAPE TOWN
PORT ELIZABETH
N

THE RACE TO THE CAPE

A story of the
UNION-CASTLE LINE
1857 - 1977

NORMAN HODSON

NAVIGATOR BOOKS

Published by Navigator Books
Ringwood, Hampshire

Copyright © Norman Hodson 1995

ISBN 0 902830 42 2

All rights reserved. No part of this publication may be reproduced, stored in a retrieval system or transmitted in any form or by any means: electronic, mechanical, photocopying, recording or otherwise, without the prior permission of the publishers

Typeset by Moorhouse Typesetters
Printed by The Cromwell Press, Melksham, Wiltshire

CONTENTS

ACKNOWLEDGEMENTS

The Author wishes to thank all those who have helped in the compilation of this story, especially "Union-Castle Chronicle 1853-1953" by Marischal Murray, published by Longman Green & Co; Mr. K. Brett for his poem "Castles" published by Adelphi Press; Mr. Sidney Smith, Group Pro, for his many publications; June Foster, Editor of the Group House Magazine "Clansman"; and the many contributors to the Magazine over the years for their anecdotes.

ILLUSTRATIONS

CHAPTER ONE

ARTHUR ANDERSON AND THE UNION LINE

For a century and a quarter the Union Line and, later, the Union-Castle Line was one of the main supporters of the port of Southampton. Even fifteen years after their services ended, wherever one went in Southampton one tended to hear of, and meet, people who had once worked for the Company and, even today, there are reunions held each year for ex-employees. The Union-Castle Line helped shape the success and importance of the City, making it one of the country's leading ports and the commercial centre of the South Coast.

It all began as a result of the careers and ambitions of two men, Arthur Anderson and Donald Currie, together with the demands, inventions and innovations of the Industrial Revolution.

At the beginning of the 19th Century, with the advent of new machinery, there was an increasing demand for coal in Industry and at the same time, in shipping, sail was giving way to steam. Coal was brought from South Wales and North East England to London and other ports by sailing barges, but freight charges were constantly increasing, the cost of rail transport was excessive and coal was generally in short supply.

In 1836 a group of businessmen met at the George and Vulture Tavern in the City of London to discuss the possibility of setting up a new docks system in Southampton. As a result, 216 acres of mud land was purchased at a cost of £5,000. Two years later, on 12th October 1938, the Foundation Stone of Southampton Docks was laid and soon afterwards the Docks were operating in what was later known as the Inner Docks.

Thereafter, because of Southampton's primary position and its geographical advantage with its unique double tides, the Docks rapidly extended. Eight years later the Inner Dock was widened and deepened and shortly afterwards another dock to be known as the Outer Dock was built with a lock between the two making it possible for the Inner Dock to be non-tidal. The first shipping companies to be attracted to the new port were the Royal Mail Line (with its trade to South America), the P & O Line (with the Far East trade) and the General Screw Steamship Company (with services to South Africa and India).

Arthur Anderson had humble beginnings from his birth in 1792 in the Shetland Islands (for which he later became M.P.). He joined the Royal Navy and served in the Napoleonic War. Afterwards he was given a job as a clerk in a small shipping office in London owned by a man named Willcox, which started his enthusiasm and interest in shipping. Willcox later made him a partner and the firm Willcox & Anderson were later appointed Managers of the P & O Line who operated their ships from Southampton.

Anderson began to realise that steam colliers could convey coal to ports more regularly and reliably than sailing barges so he formed the Southampton Steam Shipping Company but, eventually, on the 7th October 1953 the Company was finally registered as the Union Steam Collier Company, for the purpose of conveying coal to Southampton for the ocean-going packets.

Five steamships were built for this purpose and named "UNION", "BRITON", "SAXON", "NORMAN" and "DANE" (they were, of course, named after a union of races, but it was little realised that in the future they would be trading with another Union - that of South Africa). The ships were built on the Thames at Poplar and Deptford and ranged from 380 to 530 tons. However, before they could enter the service Great Britain was at war with Russia and the Government ensured that ample coal supplies were available at all ports. The purpose of the Company thus disappeared and the ships were commandeered to convey troops and equipment to the Crimea.

On 5th December 1856 the name of the Company was changed once more, this time to the Union Steam Ship Company Limited with registered offices in Oriental Place, Canute Road, Southampton. Three of the Directors were well-known Southampton men - Thomas Hull (who became Southampton Agent for the Company), Edward Dixon and Alfred Giles, who later came to more prominence. A new service to Brazil was started, but this proved too costly and was abandoned.

The original inhabitants of South Africa were the Bushmen and the Hottentots, but gradually tribes from the North moved down, especially the Bantu and Zulus and the Bushmen were pushed out into the wilds. The first Europeans to discover the Cape were the Portuguese, when Admiral Bartholomeo Diaz called there in 1488. However, they did not settle and it was not until 1652, when Jan van Riebeeck arrived in the "DROMEDARIS", that Europeans settled there, the purpose being to establish a settlement for the East India Company. Van Riebeeck was Dutch, but the rest of his men were professional soldiers and sailors - mainly German and Dutch with a number of Danes and other Europeans.

Their purpose was to establish a stronghold and refreshment station, including fresh provisions and water, for the Company's ships on the long journey from Holland to the Far East. Van Riebeeck built a fort and a number of small inns grew up around the fort to provide food, drink and entertainment for the sailors and Cape Town earned the name "Tavern of the Seas". Van Riebeeck was also said to have planted the first vines in the Cape.

South Africa became one of the most popular Colonies of the then British Empire, but of the four main territories which eventually comprised the country - Orange Free State, Transvaal, Natal and Cape Province - the latter two were the most important and populated. Over the years, many British families had settled there and they often lacked the latest news from home.

The British Government therefore, decided to provide a regular shipping service to South Africa, subject to a periodic contract, to send mail and newspapers etc. to the Cape for the British residents. In 1851 the first such contract was awarded by the Admiralty (who acted on behalf of the Post Office and who also inspected and surveyed the ships before sailing). It was granted to the Southampton based General Screw Steam Shipping Company whose first ship to sail from Southampton was the "BOSPHURUS". However, within a few years the firm collapsed. A British M.P., Mr. W.S. Lindsay, was awarded a second contract provided his ships could complete the journey to Cape Town in thirty-six days. In fact, he took fifty days and the service only lasted a few months.

PHOEBE

The Admiralty then invited tenders for a third contract and the Union Steam Ship Company submitted a tender. This was accepted on the 4th September 1857 stipulating a monthly service with a passage time of forty-two days and mails to be loaded at Devonport in Plymouth. The contract was for five years and carried an annual subsidy of £33,000. It also allowed for calls at St. Helena and Ascension Islands on the return journey from the Cape. The Company immediately brought their 530 ton "DANE" from Birkenhead to Southampton where she was dry-docked, painted, stored and loaded, but due to the shortage of time, she only carried a few passengers and the cargo on board only valued £102. On Tuesday 15th September 1857 she sailed down Southampton Water en route to Devonport and south Africa. Because of strong head-winds the journey took forty-four days, the highest speed being 7½ knots. The historic importance of that journey was obviously not realised at that time, but it was the start of 120 years of shipping between Southampton and South Africa in which thousands of journeys would be made to and from the Cape.

On arrival in Cape Town, she was given a tremendous welcome by the local people who had begun to think that everyone at home had forgotten them. Her Commander, Captain William Strutt, was entertained by many of the local dignitaries.

On her return, "DANE" shipped the largest amount of mail which up to then had been sent home from the Cape - 10,867 letters and 3,671 newspapers. She carried twelve passengers and cargo to the value of £709. The journey took only thirty-eight days, but the round voyage meant a fuel consumption of 734 tons of coal.

In the meantime, however, following the hurried departure of the "DANE", the Union Line were organising themselves for the new service. The "BRITON" had foundered off the coast of Spain and the "UNION" and the "SAXON" were unsuitable for the run and were sold. To replace them the "ATHENS", "PHOEBE" and "CELT" were purchased, the latter being the second ship to arrive at the Cape, one month after the "DANE". The "CELT" only lasted in the Company's service for five years before being sold (but not finally scrapped until 1933).

In June 1859, the Cape Legislative Assembly decided to pay the Union Line a bonus of £250 a day for every time a mail steamer completed the outward journey in less than thirty-five days.

Life on board for passengers was not easy although the adverts pointed out that cabins were furnished, since on some sailing ships passengers had to provide their own beds and bedding etc. The First Class cabins were aft, well away from the heat of the boilers etc. which the passengers in the Second Class had to suffer. When the Third Class or Steerage passengers were carried they were accommodated in the bows. For bathing, a tub was filled with salt water for gentlemen only on the open deck in the early mornings. The Company had no jurisdiction over the supply of food which was left in the hands of the Captain. The dining saloons were small and claustrophobic and were situated in the centre of the ship with cabins on either side.

However, in 1860, the first ship specifically designed for the Cape route, the 1,055 ton "CAMBRIAN", sailed from Southampton with Captain Strutt in command. She was the first ship on the route to exceed 1,000 tons and also advertised to have a bathroom for the use of passengers. She cost £25,000, had a speed of 10 knots and had accommodation for sixty First Class passengers and forty Third Class. The Mail Contract was renewed in March 1863 for seven years, reducing the passage time to thirty-eight days, but also with a reduced subsidy. Three new mail ships were ordered, the first - the "BRITON" of 1,164 tons - was claimed by her builders, Lungleys of Deptford, to be unsinkable. Shortly afterwards, the 1,142 ton "SAXON" sailed on her maiden voyage and broke another record by arriving in Cape Town in thirty-one days. In 1864, her sister ship, the 1,282 ton "ROMAN" sailed from Southampton and continued her voyage to Port Elizabeth which then became the terminal port for the Mail Service, having already been linked to Cape Town by the new electric telegraph. The "ROMAN" was given a warm welcome in the port, with much entertainment on board and a cruise around Algoa Bay for the leading citizens.

Following a Mail Contract between the Union Line and the Natal Government the Company started a new service, with smaller ships, between Cape Town and Durban for the transfer of passengers and mail. Durban was a port with a fine harbour, but which, unfortunately, had a sand bank outside called the Bar, which prohibited ships of larger draughts from entering the harbour. The "ANGLIAN" was the first ship used and was later joined by the "DANE" and the "PHOEBE", now retired from the Mail Service.

The first competition was introduced by the Diamond Line in 1864 with ships sailing from Falmouth two days before the Union Liners left Devonport. This created some interesting races between the two Companies to be the first to arrive in Cape town with news from England. This culminated in March 1865 with the "SAXON" completely overtaking the Diamond Line's "UITENHAGE" and reducing the record passage time to twenty-eight and a half days. Two years later the Diamond Line ceased to operate. The "ANGLIAN" together with the "ATHENS" started a new service to Mauritius, but unfortunately in May 1865 the "ATHENS" was sunk in high seas off Cape Town.

In 1869, a group of businessmen, including George H. Payne - a leading London ship-broker who had been involved in Cape shipping for many years - formed their own shipping company called the Cape and Natal Steam Navigation Company, for which they chartered various vessels. In April 1871, all records were broken when their ship "SWEDEN" sailed from Dartmouth and arrived in Cape Town in twenty-seven and a half days. Not to be outdone, the Union Line purchased another vessel, the 1,959 ton "SYRIA" which in October 1871 lowered the journey time to twenty-six and a half days. She was described by one South African journalist as being the finest steamship seen in Table Bay with the exception of the "GREAT EASTERN".

Still trying to compete with the Union Line, George Payne then approached Donald Currie for help and chartered two of his steamers from his Hamburg run, the "ICELAND" and the "GOTHLAND". However, before these vessels even arrived in Cape Town, George Payne's company foundered leaving the Union Line, for the time being, without competition. At the same time, the "SAXON", "ROMAN" and the "CELT" were lengthened to increase accommodation.

In Cape Town, the impending arrival of a Union Mail Liner was indicated by the hoisting of a flag on Signal Hill (a look-out post above the port, now the take-off point for the city's hang gliders). No matter what time of day this occurred, huge crowds gathered on the quayside to watch the ship berth and to receive the latest news from England. Again, no matter what the hour, there would be a postal delivery in the town.

The arrival at Southampton was not quite so dramatic, but the Company's flag was hoisted over their offices in Canute Road as soon as the news was received that the inward-bound ship had passed Hurst Castle, opposite the Isle of Wight. Even in those days, before Government Legislation, the Union Line always mustered their officers and crew for fire and boat drill on the day before sailing - all properly dressed in uniform and watched over by a Director of the Company, Mr. W.S. Nicholson.

In 1869, in addition to the monthly mail vessels, the Company decided to sail extra ships to introduce a twice-monthly service. At first, this appeared to be unfortunate timing since South Africa shortly suffered from a severe depression, no further extensions to the railway system were made and many workmen were unemployed. In addition, came the news of the opening of the Suez Canal, casting doubts on the future trade to South Africa. However, Union Line's actions and forward planning were soon to be rewarded.

Diamonds had first been discovered in South Africa in 1867, but it was two years later that the famous "Star of South Africa" was found and when it was sold for £11,000 it caught the imagination of future prospectors and the Diamond Rush was started. Soon, all Union Line ships, including their sailing ships, were full of passengers anxious to seek their fortunes. At the same time, mine machinery, railway equipment etc were being ordered from England and needed cargo space.

Arthur Anderson, the Founder of the Line, died on 28th February 1868 at the age of seventy-eight and Sir Benjamin Phillips, a former Lord Mayor of London, was appointed Chairman.

South Africa's future prosperity was secured for the time being. The Union Line was also successful as a result of this increased trade, but it was about to face even greater competition.

CHAPTER TWO

DONALD CURRIE AND THE CASTLE LINE

Born in Greenock in 1825, but brought up in Castle Place, Belfast, Donald Currie developed an early fascination for the sea and ships. At the age of seventeen he joined the newly formed Cunard Line in Liverpool (where his brother James was an engineer) and seven years later, in 1849, he opened offices in Paris and Le Havre, later relieved as Manager by his younger brother, John Martin Currie, (for many years J.M. Currie were Union-Castle agents at these places).

Donald Currie returned to Liverpool and by 1854 was virtually in charge of the Liverpool office of Cunard, but he still yearned for his own ships. In 1862, he left the Cunard Line (with no acrimony on either side, although he did agree that, in the future, he would not trade on the North Atlantic route) and formed his own company, Donald Currie & Company with a Head Office in Liverpool and with another brother, David, as a partner. He acquired an interest in a Scottish shipping company at Leith - the Leith, Hull and Hamburg Steam Packet Company Limited with steamers trading to North German ports, Scandinavia and the Baltic. Donald Currie was one of the largest Shareholders and his brother, James became Manager. Donald Currie eventually completely took over the firm.

However, he still aimed to own his own fleet and decided to trade to India which meant building sailing ships since fuel consumption was too great for steamers sailing to India via the Cape. His first four ships were built on the Clyde in 1863 and named after Castles in the British Isles - "STIRLING", "WARWICK", "ROSLIN", and "PEMBROKE". They were iron-built of 12,000 tons with attractive lines. When, four years later, the "CARNARVON CASTLE" was built she was described as one of the most graceful sailing ships of her time. In 1888, the "TANTALLON CASTLE" sailed and broke the record for sailings to Calcutta, arriving in eighty days and returning in seventy-eight, both journeys round the Cape. At that time, many ships were often several days late in leaving port after the advertised sailing times whilst their owners waited for more cargo to arrive. Donald Currie decided that his ships would sail on time whether full of cargo or not and their sailings were so advertised. In the main, traders appreciated this, knowing their goods should then arrive on time.

In 1865, he moved his Head Office to London, still keeping on his Liverpool office and transferred his ships to the Thames. Once in London, he soon became well known in the City and made many various contacts. He also extended the Leith-Hamburg run and formed the Liverpool and Hamburg Steam Ship Company with Donald Currie & Company as Managers. The ships travelled from Liverpool to Northern Europe via the English Channel.

Two events then took place which altered his plans. The first was the invention by Alfred Holt of a new type of steam engine which would enable ships to make longer voyages under steam. The second event was the opening of the Suez Canal. Currie had always wanted steam ships on a deep-sea run and the two events enabled him to build such ships for the India route via the Suez Canal. The first of these, the "DOVER CASTLE" was ordered from Barclay, Curle & Company on the Clyde.

When George Payne's Cape and Natal Company found that they could not continue trading after chartering two of his ships Currie decided to carry on running them himself, being encouraged by many merchants who thought competition on the Cape run would be good for everyone concerned. So Donald Currie entered the Cape trade, appointed Agents in Cape Town and started the intense rivalry between the Union Line and the Castle Line. He then chartered a steamer called the "PENGUIN" which caused much interest in Cape Town by arriving in twenty-four days and eighteen hours. Currie intended to sail the new "DOVER CASTLE" to South Africa, but she was diverted to another route so the first of his own steamers was the "WESTMORELAND" from the Liverpool-Hamburg Line, sailing in 1872. Donald Currie's ships sailed to South Africa from Dartmouth three days before the Union Mail Liner and often arrived in the Cape up to ten days before them. En route they called at Bordeaux and

Currie made an arrangement with Reuters to telegraph the latest news from London to the ships at that port which then reached the Cape quicker than the official news with the Union Line.

Shortly afterwards he moved his Head Office to larger premises in Fenchurch Street and a few years later formed a limited public company called the Castle Mail Packet Company Limited. In 1880 he was elected M.P. for Perthshire and the following year the Queen bestowed on him a Knighthood.

Vessels of the Union Line were first handled in Cape Town by the agency of Wm. Anderson & Company until the Line opened their own offices at 16 Adderley Street in 1876.

The Castle Line appointed as their agents a Captain James Murison together with his partner Mr. William George Anderson (no relation to the Union Line's agent or to the Founder of the Line). Their office was also in Adderley Street on the site of which the later Union-Castle offices were built. In the later stages of the Union-Castle Line, they had offices at Cape Town, Port Elizabeth, East London, Durban, Johannesburg, Salisbury, Lourenco Marques, Dar es Salaam and Mombasa.

CHAPTER THREE

THE RACE TO THE CAPE

In 1872 the Post Office was ready to agree a new Mail Contract with the Union Line providing a twice-monthly service in a stipulated thirty days with an additional steamer once a month. However, for various reasons this was not ratified by the British Parliament so the existing contract continued until 1876.

In 1872 the first Castle appeared in Cape Town - the 2,446 ton "WALMER CASTLE", originally intended for the India route. The following May the journey time of the outward trip was again reduced, this time to twenty-three days by the 2,672 ton "WINDSOR CASTLE" which the previous September had made a record breaking journey to Calcutta before being switched to the Cape. Donald Currie was aiming to receive a share of the Mail Contract and in 1875 built a ship specifically for the run, the 2,880 ton "DUNROBIN CASTLE". She carried 100 passengers and, for the first time, the First Class Dining Saloon extended the full width of the ship and on her maiden voyage immediately reduced the voyage time by twelve hours.

The Union Line's contract still had three years more to run and the Cape Government, keen to keep the Castle Line in competition, offered Donald Currie £150 a day for every day his ships completed the outward journey in less than thirty days. Further negotiations over the contract took two years including talks in London with the Prime Minister of the Cape Government. Pressure was being brought in both countries for both Companies to be included in future contracts. However, due to a previous disagreement with the Cape Government over a possible East African service, the UK Government indicated that they did not wish to be further involved in Mail Contracts to South Africa. In 1876 therefore, the Cape Government issued their own contract stipulating that it would be shared equally by both the Union Line and the Castle Line with a weekly Mail Service and alternate sailings by each Company. The journey time was to be twenty-six days. There was also more postage money available for the mail carried, but an important clause was that the two Companies were not to merge. A similar contract was given by the Natal Government for the carrying of mail from Cape Town to Durban and both Companies produced fleets of smaller ships for this purpose.

Durban was first discovered by Vasco Da Gama on one of his voyages of discovery and since he landed on Christmas Day 1497 he called it Port Natal. Eventually a group of British settlers made their homes there and in 1835 it was renamed Durban after Sir Benjamin D'Urban, the then Governor of the Cape Colony. By 1860 the area around Durban was covered in sugar cane plantations with mainly Indian workers.

With ships bringing passengers and mail from Cape Town the port developed, but was rather restricted by a sand-bar at the entrance of the harbour, but by 1888 had become the terminal port of the Mail Service. Passengers and cargo were handled outside the harbour and, in 1880, the Princess Eugenie of Russia was one of many passengers who were landed over the side of the Union Line's "GERMAN" in large baskets on to the deck of the tender.

By 1904 breakwaters had at last been built in the harbour and ships could at last cross the Bar and dock alongside, the first being the "ARMADALE CASTLE". Since then Durban has grown into the country's third largest city and the country's busiest port.

The division of the Mail Contract did nothing to lessen the rivalry, but a period of reorganisation took place. The Union Line moved their Head Office from Southampton to Leadenhall Street in London and announced their intention of moving their ships to the Thames. This provoked a massive outcry in Southampton where the majority of the crews lived, as did many of the Shareholders and where so many of the traders depended on the Line. Only one Director, Alfred Giles, voted against the proposal. There was a rush to buy Shares in the Company in an effort to influence the voting and, in the end, the Southampton factor won. All the Directors resigned except Alfred Giles who was appointed Chairman.

Alfred Giles did much to enhance Southampton's reputation as a port. He was M.P. for the town for twelve years and Port Harbour Master and had been responsible for the construction and improvement of the Docks including the building of the Empress Dock. He resigned through ill-health in 1895 and died the following year. The new Chairman of the Union Line was Sir Francis Evans, a member of the Southampton Dock Company (later taken over by the London and South West Railway) and a Liberal M.P. for Southampton.

Another Director appointed to the Board of the Union Line was G.W. Wolff M.P., a partner in the Belfast shipyard of Harland & Wolff which then built all new vessels for the Union Line and for many years those of the Union-Castle Line. The designer of all these vessels was William Pirrie (later Lord Pirrie, Chairman of Harland & Wolff and Vice Chairman of Union-Castle Line).

Another result of the division of the contract was that the Union Line, now only providing twenty six voyages per year instead of the former thirty six, had ships unemployed so they started what was called an Intermediate Service from Southampton to Port Elizabeth (passengers and cargo only - no mail). By 1881 these sailings were extended to once a fortnight so that in conjunction with their Mail Service the Union Line had a weekly sailing from Southampton. Once a month, these ships also called at Antwerp and Hamburg.

Trade was booming and both Companies ordered new tonnage: - "PRETORIA", "ARAB", "TROJAN" and "SPARTAN" for the Union Line, and, - "KINFAUNS", "GARTH", "DRUMMOND" and "GRANTULLY CASTLES" for the Castle Line.

More competition came onto the scene, mainly from the Clan Line, Rennie's Aberdeen Line and the Bullard King Line. Ships of the latter two companies were smaller and could dock alongside in Durban. In 1883 Sir Donald Currie called a meeting of the various Companies involved and formed the first South African Shipping Conference to decide on freight rates and other joint problems.

Sir Donald had a large estate at Garth in Perthshire (for which he was now M.P.) and where he often entertained the Prime Minister - W.E. Gladstone. His next move was to launch his largest ship yet, the 4,300 ton "HAWARDEN CASTLE", named after Gladstone's home in Wales and launched by Mrs. Gladstone.

Shortly afterwards, Currie bought a 4,000 ton steamer which he named "PEMBROKE CASTLE" and from which he obtained the ultimate in publicity. Before she sailed on her maiden voyage he arranged a cruise from the Clyde around the North of Scotland to which he invited the Prime Minister - Mr. Gladstone; the Poet Laureate - Lord Tennyson; and many other distinguished guests. Whilst off Cape Wrath, it was decided to extend the journey to see the coast of Norway and to visit Denmark. As it happened, a number of European Royal Families were gathered in Copenhagen, including the Prince and Princess of Wales, the King and Queen of Greece, the Czar and Czarina of all Russia, the King and Queen of Denmark and the Royal Family of Hanover. They invited the Prime Minister and Sir Donald to join them for dinner and later were invited back to a reception on board the "PEMBROKE CASTLE" - twenty-nine Royals, still today probably the most eminent gathering on board a British merchant ship. The news of the event was telegraphed by the press and by this time Mr. Gladstone realised that he had not obtained the necessary constitutional permission of the Sovereign for a Prime Minister to leave the country. As may be expected Queen Victoria was not amused, but after several exchanges of letters all apparently was forgiven, but the Queen admitted that the barony she was thinking of bestowing on him would be postponed.

Two years later there was another Royal occasion when Queen Victoria's grandson Kaiser Willhelm of Germany was to open the Kiel Canal. Sir Donald Currie arranged a cruise on board "TANTALLON CASTLE" to which Mr. Gladstone was again invited although by then he was no longer Prime Minister. The ship sailed to Hamburg and then to Copenhagen where the King and Queen of Denmark lunched on board before the ship proceed to Kiel for the opening ceremony.

In 1883 another Mail Contract was signed with the Cape Government stipulating twenty-one and a half days for the journey with calls at Madeira and St. Helena. Penalties were fixed if the time took more than twenty-three days and, instead of a percentage of the postage, a fixed subsidy of £25,000 was paid to each Company with the usual stipulation that the two Companies were not to merge. The Mail Contract of 1888 further reduced the journey time to twenty days.

From 1883 South Africa came under another period of severe depression and several of the extra vessels were laid up. The fortunes

SCOT 1891

of the Mail Service, however, were saved (as they had been during the Diamond Rush) by the discovery of gold in the Rand. The Gold Rush brought both Companies back to full capacity, but no new tonnage was ordered for several years. The Union Line's "MOOR", however, was sent to Belfast for a refit. There she was lengthened by fifty-five feet, her original tall funnel replaced with two (one a dummy) and also re-engined. She was also the first ship of the Fleet to be fitted throughout with electricity - previously it had only been available in the Dining Saloons.

In 1897 two more Royal occasions followed. First, Queen Victoria celebrated her Diamond Jubilee with a Naval review at Spithead, at which were present the "ARUNDEL CASTLE", "NORHAM CASTLE", "GRANTULLY CASTLE", "PRETORIA", "ASIATIC" and "MOOR". The next occasion in the same year was when the Prince of Wales started a grand yacht race from the deck of the 4,241 ton "NORHAM CASTLE" off Southend before going on a short cruise on the liner.

Fierce rivalry between the two Companies continued. In 1890 Currie ordered an even larger ship, the 5,625 ton "DUNNOTTAR CASTLE", with two funnels and three masts, capable of carrying 360 passengers. On her trials she reached a speed of 17½ knots. Again, before her maiden voyage, she took a party of Company guests for a cruise around Scotland. There was no doubt that Sir Donald Currie had been accepted in High Society and had a good sense of publicity. After the cruise the "DUNNOTTAR CASTLE" picked up the mail at Dartmouth and reached Cape Town in just seventeen days and twenty hours.

Not to be out-done, Alfred Giles ordered a new Union mail ship to be built on the Clyde and called it the "SCOT" - 6,884 tons with twin screws and clipper bows with an elliptical stern. She was, by common consent, a very handsome ship and was one of the last to have a carved effigy at the bow.

To get one better Sir Donald announced that, after negotiations with the Post Office, all future Castle mail steamers (who still originally sailed from the Thames) would load the mail in Southampton, thus cutting out the call at Dartmouth. The first such sailing was that of the "DUNNOTTAR CASTLE" and at a reception on board, just after sailing when he entertained the Mayor of Southampton and other local officials, Sir Donald forecast that it was the beginning of a new era for the port of Southampton. Travelling on board were the British Rugby team to tour South Africa and Sir Donald Currie presented the captain of the team with a gold cup to be competed for in South Africa and which became known as the Currie Cup.

In July 1891 the "SCOT" left Southampton for the first time and reached Cape Town in fifteen days and nine hours including a call at Madeira. The "SCOT", however, proved very expensive to maintain and did not help the finances of her owners at that time. Nevertheless two years later, in March 1893, she arrived in the Cape three days early taking only fourteen days and eighteen hours to complete the outward journey. This was a record that was to stand for forty-three years since the following year the two Companies had agreed to discontinue 'racing' - not only was it uneconomical, but the general opinion in south Africa was that it was preferable for the mail vessel to arrive on a regular day.

The competition on the Cape run was then joined by the introduction of a new German shipping company who later joined the South African Shipping Conference. Equipment for the Rand mines was not only being imported from the UK and the Continent, but also from North America so both the Union Line and the Castle Line commenced new services from USA to South Africa. These services continued for many years and the last ship to arrive from that country was the 5,239 ton cargo vessel "DUNBAR CASTLE" in 1939 (she was later lost by fire in the Red Sea in 1943).

By the early 1890's the Harland & Wolff influence on the Union Line came to the forefront. Their chief designer, W.J. Pirrie, built three new Intermediate ships called the 'G' Class ("GAUL", "GOTH" and "GREEK" - there were later ten of these ships) with more cargo space, but of a lighter draught. Harland & Wolff then built a new mail ship for the Union Line based on their experience of building ships for the North American route. The "NORMAN" of 7,500 tons had accommodation far superior to any other vessel on the Cape run and for the first time First Class passengers were accommodated amidships instead of at the stern. She had a somewhat chequered career being placed in reserve at Netley in 1910, did trooping duties at the beginning of the War in 1914 before returning to the mail run. In 1918 the "NORMAN" was back on trooping duties in the Mediterranean, then on the mail run until 1923 when she was switched to the Round Africa run before finally being scrapped in 1925.

Both Companies were trading well, but little cargo was leaving South Africa except gold and diamonds. Bigger ships were being built including Union Line's "BRITON" of 10,000 tons and Castle Line's "KINFAUNS CASTLE" and "KILDONAN CASTLE" and later Union Line's "SAXON" of 12,000 tons. More attention was being paid to passenger accommodation and facilities.

However, the parallel existence of the two separate Companies was drawing to a close. Sir Donald Currie had long wanted to take over the Union Line and in the end he was unwittingly assisted by the Cape Government who wanted a still better service. When the Mail Contract was due for renewal in 1900 they announced it would be given to one company only. Neither of them tendered, mindful of the extra tonnage required by the successful firm and the surplus tonnage by the loser. Eventually it was divided as before, but significantly the non-merger clause was omitted.

CHAPTER FOUR

EAST AFRICAN AND OTHER ROUTES

Before the mid-19th Century East Africa was largely unknown to the Western World. It took the arrival of Stanley Livingstone in 1856, after his epic journey across the Continent from West to East, to bring the area to the attention of the public. Even then he concentrated mainly on a campaign to abolish the flourishing slave trade which was centred on the island of Zanzibar.

The Northern part of the area was in the hands of the Arabs who were running the slavery whereas the South was mainly Portuguese territory. Following Livingstone's plea to stop the trade, the point of view was expressed in London that if a steamship service called regularly the natives would realise that other countries and people existed other than the Arabs and it could open more opportunity for them than slavery.

The British Government became interested and the Chancellor of the Exchequer, Mr. Robert Lowe, asked the Post Office to see if the Union Line were prepared to run two services to Zanzibar - one from the Cape and one from Aden - at both these ports to connect with shipping companies on other routes. At first they were reluctant to take on this new commitment due to possible navigational problems and an expected lack of trade. However, they eventually agreed they could run a service from the Cape at monthly intervals and at an agreed fixed subsidy. They also wanted the existing South African Mail Contract extended for another three and a half years (making a total of eight years) and to be allowed to sail a third steamer to the Cape each month. British India Company were also interested, especially the route from Aden.

This resulted the following year by Mr. Lowe authorising the Union Line to run both routes and increase the frequency of the Mail Service. Union Line promptly increased their fleet by purchasing two small ships for East Africa, the "BASUTO" and the "ZULU" as well as four new mail ships, "ANGLIAN", "AMERICAN", "TEUTON" and the "NYANZA". All this was done without the knowledge of the South African Government and no mention had been made of the need for the Agreement to be ratified by the British Parliament. Following rumours which were being circulated in the City of London and many protests in South Africa (possibly inspired by Donald Currie whose Castles were now appearing in the Cape), the contracts were laid before Parliament. The enlargement and extension of the Mail Contract were postponed, but the Union Line was eventually awarded an eight year contract for the Cape-Zanzibar route which started in 1873.

East Africa was then developing very quickly and the Germans reached an agreement with the local chiefs resulting in a Protectorate of German East Africa. In the mid 1880's Britain was granted a Protectorate over British East Africa and authority and right of exploitation was granted to a private company - the Imperial British East Africa Company.

They sent out various expeditions to explore the interior, but were met with considerable hostility from the various tribes and eventually the Company could find very few resources to help them govern the country and to start trading. In 1895 therefore, the British Government itself took over the administration of the area which later became known as Kenya and Uganda. A railway line was started inland from Mombasa which eventually reached Nairobi in 1899 and Lake Victoria two years later. White settlers were encouraged to farm on land adjacent to the railway which started one of the country's major industries.

Mombasa itself had had a chequered history having been invaded and occupied by many nations, especially the Arabs and Portuguese and for some years the Sultan of Zanzibar governed the port. Roads inland were also built together with a railway line to Uganda for almost the entire trade for that country came through Mombasa. The building of this line virtually brought to an end the practice of slave trading. Many companies began to move their headquarters to Nairobi and later the Government itself moved there.

After the First World War, in 1920, many of the German possessions in Africa were disposed of, that of the Protectorate of East

LLANGIBBY CASTLE

Africa was given to the British and called Tanganyika. The Protectorate of British East Africa was annexed and renamed Kenya and Uganda. The German South West Africa was given to South Africa and shortly afterwards Union-Castle liners called regularly at Walvis Bay. In 1926 the first deep-water dock was opened in Mombasa and the first Union-Castle ship to use it was the cargo ship "CRAWFORD CASTLE". The Zanzibar service did not survive nor did a British India run from Aden - mainly due to lack of trade. The Germans were now running the only regular service to this part of East Africa and the British were dependent on them for supplies to their territories. The then Union-Castle Line had been running a service from UK to Lourenco Marques via the Cape and in January 1910 they extended this to include Beira and Mombasa commencing with the "DUNLUCE CASTLE" and followed at four week intervals by two other 'D' Class ships, the "DOVER CASTLE" and the "DURHAM CASTLE".

The Company wanted to start another UK-Mombasa service via the Suez Canal, but the Canal dues were considered too high and the Government refused a direct subsidy but eventually guaranteed all Government passages and freight to be sent out this way. This resulted in the sailing from London of the 4,900 ton "GUELPH" in what was described as Royal East African Service, calling at Mediterranean ports, Aden and Mombasa before continuing down to Durban to connect with the mail steamers (the service from Lourenco Marques being discontinued). In 1917 Union-Castle Line opened their own offices in Mombasa. Larger ships were gradually used on the route, such as the 6,000 ton "DUNVEGAN CASTLE" and the "CARISBROOK CASTLE", until they were scrapped in 1922, culminating in 1914 with the 11,200 ton "LLANDOVERY CASTLE" and the "LLANSTEPHAN CASTLE", both of which proved very popular with passengers.

During the First World War the service was suspended but was restored again in 1922 with some vessels returning via the East coast and others sailing out West and returning home East. Thus started the Round Africa Service which continued successfully for many years, later with more modern and larger ships such as new 'L' Class ships (including the "LLANGIBBY CASTLE"), "DURBAN CASTLE",

"DUNNOTTAR CASTLE" and "WARWICK CASTLE". The last ships on this route were, of course, "BLOEMFONTEIN CASTLE", "BRAEMAR CASTLE", "KENYA CASTLE" and "RHODESIA CASTLE".

Until 1890 Beira was just open wasteland - there were no buildings, just the mouth of the Pungwe River. It started to develop in the late 1890's and was used to open up the new country of Matabeleland - later to be known as Rhodesia named after Cecil Rhodes, a former Governor of the Cape.

Journeys inland were by means of the river, but eventually roads and railways were built. Both the Union Line and the Castle Line played leading roles in the development, although there was fierce competition from Bullard King and German lines. The Union Line's "NORSEMAN" and "TYRIAN" together with Castle Line's "COURLAND" and "VENICE" sailed alternately from Durban with supplies and equipment for the opening up of the upper reaches of the river which proved no easy task and it was often necessary for steamers' surgeons to provide medical treatment.

In April 1909 the "YORK CASTLE" sailed from Middlesborough for Beira loaded with railway equipment. Until a railway line was built from Beira to Blantyre in the 1920's, Union-Castle Intermediate ships called at Chinde in Nyasaland to serve the needs of that country. In 1915 the Company purchased a small 720 ton coaster, the "IPU", for service between Beira and local ports. She was later replaced by the 1,300 ton "ROVUMA".

South of Beira, still in Portuguese East Africa (later called Mozambique) is the port of Lourenco Marques, in Delagoa Bay. Apart from a natural harbour, the interior was just bush land largely occupied by wild animals. The port developed as a result of a decision in 1887 to establish a railway line into the Eastern Transvaal to bring in equipment for mining the newly found gold. Although the prospectors landed at Cape Town and had to make their long and laborious way up country, it was considered easier to transport the heavy equipment via the East Coast.

The first Union Line vessel to call was the coaster "KAFFIR" from Durban, but in May 1887 the "DUNBAR CASTLE" sailed from

London to bring out the first consignment of mining and railway equipment from the UK. The line was eventually opened in 1894.

Both Union Line and Castle Line Intermediate ships called at Lourenco Marques and, to encourage Dutch settlers, called in at Holland en route - Union ships to Rotterdam and Castle ships to Flushing. The Portuguese were reluctant to use their own ships to their East African territories so British ships conveyed their passengers with some of them calling at Lisbon. On this route too, the Germans provided competition and offered cheap through-fares from Germany to the gold fields in Transvaal.

In 1890 Donald Currie ordered two new ships for the route, the 4,045 ton "DOUANE CASTLE" and the "LISMORE CASTLE". Although the Union Line had often used the term these were the first two ships which Currie called 'Intermediate'.

When the Boer War started in earnest, many British miners and other refugees flooded out to the neutral territory of Lourenco Marques where they embarked for the Cape and England.

In 1864 the Union Line established another new service carrying mail to Mauritius since the island had become a general focal point with many other shipping companies serving the Far East. UK bound mail sent on this route was obviously more expensive, but due to various spaced sailing days it often reached England several days before that brought in the mail steamer. The service from Cape Town, Port Elizabeth and Durban to Mauritius was maintained by the "ATHENS" (until she was lost in high seas off Cape Town), "ANGLIAN" and "MAURITIUS". On arrival at the island the mail was trans-shipped to a P & O steamer bound for the UK via Suez. It was off loaded at Marseilles and continued its journey to England overland and thus became known as the Overland Mail Service.

Trouble, however, was experienced when rough weather at Durban often caused delays of several days and often missed the P & O connection. Gradually the port of Galle in Ceylon became the junction of Far East shipping but, although the Company extended its Mauritius service to Galle, it did not last long and both services were discontinued.

CHAPTER FIVE

THE UNION-CASTLE LINE

With the omission of the 'no merger' clause in the new contract, Sir Donald Currie immediately suggested forming an amalgamated Company with Donald Currie & Company as Managers (as they always had been of the Castle Line). He offered a partnership in Donald Currie & Company to Sir Francis Evans, Chairman of the Union Line. The new Company - to be called the Union-Castle Mail Steamship Company Limited - would take over all profits, liabilities and assets of the Union Line together with its fleet, forming a new shipping company with some 221,000 tons of shipping. The existing Directors of the Union Line would become Directors of the new Company and all employees of the Union Line would be given fair treatment. This was agreed by Shareholders of both Companies and, on the 8th March 1900, the new Company was so registered.

The staff of the Union Line were obviously sad to hear the news, but accepted that they had been fairly treated. Two days later, on the 10th March, the "MOOR" sailed from Southampton for the last time in Union Line colours (yellow funnel and white hull). Two weeks later, the "BRITON" sailed in the new Union-Castle colours, the famous lavender hull with red and black funnels.

In the meantime, however, on the 17th March Sir Donald Currie hosted a gathering on the promenade deck of the "DUNNOTTAR CASTLE" which included the Duchess of Teck and Lady Roberts, wife of the future C. in C. South Africa (both ladies were travelling in the ship). The party witnessed the hoisting of the Union-Castle house flag, a skilful combination of the two previous house flags. The original Castle blue was included, with the central 'C' omitted while on the white cross of St. Andrew was superimposed the Union line's red diagonal.

Shortly before the amalgamation, the Boer War started in South Africa and many Intermediate ships were requisitioned by the Government for carrying troops and equipment. Even as the new flag was being broken out on the "DUNNOTTAR CASTLE", the "GAIKA" (ex Union Line) moored astern and ready to sail laden with troops, also broke out her new flag.

The "SPARTAN" and the "TROJAN" were converted into hospital ships and the new "KILDONAN CASTLE", completing on the Clyde, was hastily converted into a troopship. It was left to the "DUNNOTTAR CASTLE" to get the most publicity in October 1899, when Sir Redvers Buller and his staff embarked to take command of the British forces in the Cape. He was received on the quayside by the Mayor and Corporation of Southampton and welcomed on board by Mr. Frederick Mirrieless, the son-in-law of Sir Donald, who was abroad at the time. The departure from Southampton was witnessed from an upper deck by a young twenty-four year old War Correspondent from the London Morning Post, Winston S. Churchill. He later described the peacefulness of the voyage, but emphasised that for two weeks they were all cut-off from any news of the War. Nevertheless the voyage must have been far different than when he sailed from Berth 38 in Southampton Old Docks after his retirement, on "DURBAN CASTLE" for a painting holiday in Madeira. The accommodation, at least, was more luxurious.

In December of that year, after defeats in South Africa, Lord Roberts embarked on "DUNNOTTAR CASTLE" to take up duties as Commander-in-Chief and his Chief of Staff, General Kitchener, embarked at Gibraltar. After the Boer War ended in 1902 more exports were expected from South Africa, but they did not materialise. Currie built for the first time four pure cargo ships - "YORK CASTLE", "GORDON CASTLE", "AROS CASTLE" and "CORFE CASTLE", but they were not needed and eventually employed on the USA-SA run. Three new Intermediate vessels of the new 'D' Class were also built, "DURHAM CASTLE", "DOVER CASTLE" and "DUNLUCE CASTLE", which proved very popular with passengers.

Following the amalgamation and the end of the War, there was obviously a surplus of shipping and at one time fifteen of their ships were laid up at Netley in Southampton Water and on the Thames. The Mail Service then comprised nine ships, they were:

"SAXON"
"BRITON"
"NORMAN"
"SCOT"
"KILDONAN CASTLE"
"KINFAUNS CASTLE"
"CARISBROOK CASTLE"
"DUNVEGAN CASTLE"
"WALMER CASTLE"

In 1901 the "TANTALLON CASTLE" had been lost, going aground near Robben Island (once the site of a famous prison) off Cape Town. The passengers were all safely ferried ashore, but despite the help of tugs the vessel would not move and eventually broke up in high seas.

The following year Union-Castle Line ordered the first mail ships for the Company - "ARMADALE CASTLE" from the Clyde and "KENILWORTH CASTLE" from Belfast, at 13,000 tons the largest yet ordered. When they came into service the "SCOT" and the "DUNVEGAN CASTLE" joined the ships in reserve at Netley. Mail ships now sailed from Southampton at 5.00pm on Saturdays and the weekly Intermediate ships alternated between a Friday departure from Southampton and Saturdays from London, including calls at Continental ports en route to the Cape.

A new Mail Contract was due in 1908 for which the Company were the only firm to tender, but the four South African Governments were more interested in the unification of their territories and asked the Union-Castle Line to continue the existing contract until 1912. However, Sir Donald ordered two new mail vessels "BALMORAL CASTLE" and the "EDINBURGH CASTLE" both of 13,000 tons and they turned out to be the last of Currie's Castles.

By then Sir Donald was eighty-two and both his sons who had trained in the work of the Company had left to seek their fortunes elsewhere. Sir Francis Evans had died in 1908. In April 1909 Sir Donald Currie died in Devon and memorial services were held in St. Paul's Cathedral in London and also in Southampton, Cape Town and other ports. Thus ended an era which proved monumental in the history of British Shipping in general and that of the port of Southampton and of South Africa in particular. His son-in-law, Mr. Frederick (later Sir Frederick) Mirrieless became Chairman of the Company.

CHAPTER SIX

THE POST-CURRIE YEARS

For some time discussions had been taking place in South Africa between the four Governments regarding the unification of the country and in July 1909 the four Prime Ministers came to London for further talks with the British Government. These were satisfactorily concluded and on the 31st May 1910 the Union of South Africa came into being. H.R.H. The Prince of Wales agreed to open the new south African Parliament in Cape Town and the Company's largest and newest liner, "BALMORAL CASTLE" was taken over by the Admiralty for the voyage to South Africa. She sailed to Portsmouth where special accommodation was prepared for the Royal Party. Her funnel and masts were painted yellow, a cross-tree fitted to her mast for the extra flags required and as "H.M.S. BALMORAL CASTLE" she flew the White Ensign. Although some engineers and stewards stayed on board, the majority of the crew were from the Royal Navy.

However, shortly before she was due to sail the death occurred of H.M. King Edward VII and the Prince of Wales ascended the throne as H.M. King George V. King Edward's brother, the Duke of Connaught, agreed to open the Parliament in his place and accordingly on 11th October 1910 he embarked at Portsmouth with the Duchess and their daughter, Lady Patricia.

A remarkable painting by Norman Wilkinson depicts the ship proudly sailing out into the Solent flying the Duke's personal standard, his flag as Master of Trinity House, the Union Flag of Great Britain and the White Ensign.

The Duke duly opened the new Parliament in November 1910 and later laid the Foundation Stone of the new University of the Cape of Good Hope before sailing to Durban and then returning home to the UK.

A new Mail Contract was due to be completed in 1912. However, negotiations involving Mr. Percy Molteno, the late Sir Donald's son-in-law and the new South African Government came to a halt. When Sir Donald Currie formed the South African Shipping Conference, they had agreed a system of rebates in respect of freight charges which did not find favour in South Africa. In 1911 their new Parliament passed a South African Post Office Act, one of the clauses of which stated that no Company applying the rebate system could be awarded the Mail Contract. When tenders were again invited no Company tendered and so the matter stood for a while.

In 1911 there was a sudden sharp rise in the price of Union-Castle shares on the Stock Exchange. It was thought by many City traders that they were being bought by Lord Pirrie, Chairman of Harland & Wolff and by then associated with Sir Owen Philipps, Chairman of the Royal Mail Line. In the December of that year it was officially announced that the Royal Mail Line and the Elder Dempster Line had jointly bought out the Union-Castle Line. One of the conditions of purchase was that the firm of Donald Currie & Company were no longer to be Managers of the Company and the seven partners of the firm, mostly relatives of Sir Donald, were all paid off. This came forty years after Sir Donald entered the South African trade.

CHAPTER SEVEN

THE ROYAL MAIL INFLUENCE AND THE FIRST WORLD WAR

Sir Owen Philipps had been brought up in shipping and later invested in some shares in a small company which he registered as the King Line and became its first Chairman. He went on to become Chairman of the Royal Mail Line (the oldest British Shipping company) and later still acquired the Shire Line, Pacific Steam Navigation Company Limited and Elder Dempster Line. With the purchase of Union-Castle Line he created one of the largest shipping groups in the world. Among the new Directors of Union-Castle Line were Lord Pirrie and Mr. A.H. Milbourne (who remained a Director for many years until his retirement) and among the three Joint Managers appointed was Mr. Robertson Fyffe Gibb, originally of the Union Line.

Immediately on taking over as Chairman, Sir Owen ordered two new 11,000 tons ships for Intermediate Service to East Africa. He also started negotiations with the South African Government for a new Mail Contract at the same time agreeing to abolish rebates. He sailed for the Cape on the "BALMORAL CASTLE" and was almost regally welcomed to South Africa by the Prime Minister, General Botha. He returned on the "WALMER CASTLE" with the Minister of Posts and during the voyage the new contract was agreed. There were to be no rebates, the outward journey to be completed in sixteen days, six new steamers to be built of not less than 15,000 tons and also new freight rates were agreed for the Northbound journey. Special arrangements were also to be made for the provision of refrigeration space on mail vessels to allow for the shipment of South African fruit to the UK. Sir Owen Philipps also undertook to take more interest and participation in the country's affairs.

However, the rebuilding programme was interrupted by the outbreak of the First World War, during which Union-Castle played a major part among the various shipping companies, especially having its base at Southampton which was so accessible to France.

Even before the formal Declaration of War, the "CARISBROOK CASTLE", "NORMAN" and "DUNVEGAN CASTLE" were taken out of the reserve at Netley and commissioned by the Admiralty, the first as a hospital ship and the latter two as troopships. Within a month, nineteen out of the forty-one ships in the Fleet has been commandeered by the Royal Navy. Within a week of the War starting, the "NORMAN" and "DUNVEGAN CASTLE" were among the ships to land the first men of the British Expeditionary Force in France.

The first two mail ships to reach Southampton were the "KINFAUNS CASTLE" and "ARMADALE CASTLE" which were immediately stripped of their passenger accommodation and refitted as Armed Merchant Cruisers. They were each equipped with eight guns and sent out to the South Atlantic where they were based at the Royal Naval Dockyard at Simonstown in South Africa. The then "H.M.S. ARMADALE CASTLE" in particular saw much action in her new roles and together with "H.M.S. KINFAUNS CASTLE" and the former Intermediate liners "GALWAY CASTLE" and "GAIKA" were involved in the capture of both German West African and German East African territories. In 1916 "H.M.S. ARMADALE CASTLE" shipped £7,000,000 in gold and specie under special orders for an unknown destination, which later proved to be Halifax. Afterwards, she helped to escort a convoy carrying 16,000 Canadian troops to Britain. After a refit in Liverpool she was engaged in blockade work between the Shetland Islands and Norway before once more acting as convoy escort between England and Sierra Leone.

The "KILDONAN CASTLE", which had served as a troopship during the Boer War, was again pressed into Royal Naval service, first as a troop ship and then, during the Dardanelles Campaign, as a hospital ship. Throughout the War the Company managed to keep a regular Mail Service running to South Africa, bringing in many of the former Intermediate vessels since this service had been temporarily abandoned. An interesting aspect of the War was that two of the former Union Line ships whose names were connected with the enemy were renamed - the "GERMAN" became the "GLENGORM CASTLE" and the "GALICIAN" renamed "GLENART CASTLE".

The next vessel to be taken over was another mail ship. Lord Kitchener wanted more troops on the Western Front and, in particular, a battalion stationed at Gibraltar. The "EDINBURGH CASTLE" was steaming Northbound with mail, cargo and a full complement of passengers. She was ordered to divert to Gibraltar where she embarked the troops who remained on deck until arrival at Southampton where they disembarked before being conveyed to France. After disembarking the passengers and discharging the cargo, the liner was also converted to an Armed Merchant Cruiser.

The First World War also meant the end of another era in South Africa. For one hundred years British troops had provided the garrisons for the four provinces, but now they were needed to fight for their own country. Six Union-Castle liners were assembled in Cape Town to embark them and sail to England escorted by two cruisers. Thereafter Union troops provided their own garrisons. The Company's ships were also involved in bringing South African troops over to the War in France, including General Smuts who came to England to join the War Cabinet.

Many of the ships were converted to hospital ships and did yeoman service in various ports of the world including France, India, the Mediterranean, East and South Africa. Two of them were sunk despite their Red Cross markings - the "GLENART CASTLE" and the "LLANDOVERY CASTLE" - the latter was torpedoed whilst en route from Halifax to Liverpool by a German submarine which later surfaced and shelled the lifeboats leaving only 24 survivors out of 258 on board. The "CARISBROOK CASTLE", "KENILWORTH CASTLE" and "DOVER CASTLE" were all hospital ships stationed at Gallipoli in 1915.

In 1917 the "KILDONAN CASTLE" sailed under sealed orders for Oban where she embarked some V.I.P.'s to take to Murmansk to form an Allied Mission to Russia in an attempt to stop the collapse of the Russian Empire. The ship remained there a month, most of the time iced in, before the Mission returned having failed in their attempt (the Russian Revolution started shortly afterwards).

CARNARVON CASTLE

Later that year, following Russia's signing of a peace treaty with Germany, another force was sent to Murmansk to prevent the Germans from establishing a submarine base in the port and to rescue a massive supply of stores which had been left there. Included in the Force was the "BRAEMAR CASTLE" which was designated a base hospital and she stayed in the port for nearly a year.

Shortly after this the first ship on the mail run was sunk - the former Intermediate liner "ALNWICK CASTLE" whilst en route from Plymouth to South Africa - however, all but forty of the crew and passengers were rescued. Altogether, eight of the Company's ships were lost during the War and six seriously damaged, with a total loss of 440 lives. After the War Union-Castle ships were engaged, along with those of other Companies, in returning troops to their homelands throughout the world from the UK and the Continent as far as Australia and New Zealand.

An interesting tail-piece is that of the return to the UK of the "CARISBROOK CASTLE" who had signed on as Trimmers in the Engine Room various men who had served in the War and had been repatriated to South Africa, but now jumped at the chance to return to England. However, the heat in the Engine Room proved too much for them and the Captain appealed for volunteers which was taken up by men of the King's African Rifles who were able to work in the heat.

ARUNDEL CASTLE

CHAPTER EIGHT

POST-WAR YEARS

It was October 1919 before the Mail Service was fully resumed with sailings at 4.00pm every Friday and 1922 for the Round Africa service which sailed from London every fortnight.

The first mail ship to be built under the direction of the Royal Mail Line was the 19,216 tons "ARUNDEL CASTLE". Launched in 1919 and delivered in 1921, she was a complete break from tradition. Apart from being the largest liner so far built for the Cape run, she had four funnels (one a dummy) and far better facilities for passengers including, for the first time, a swimming pool on the upper deck.

In 1924 she was the first ship to use the new 3 acre 7,500 ton floating dry dock which was permanently moored off the quayside in Southampton Old Docks. It was officially opened by H.R.H. The Prince of Wales which he did by sailing into it on the steamer "DUCHESS OF FIFE". With the closure of Southampton docks in the 1940's the dry-dock was towed to Portsmouth where it remained for many years. In 1985, however, as it was being towed to Brazil it split in two and ended up as scrap.

Her sister ship, the "WINDSOR CASTLE" was launched on the Clyde in 1921 by H.R.H. The Prince of Wales - the first member of the Royal Family to launch a merchant ship.

Apart from his many shipping interests, Sir Owen Philipps also had connections in the City of London and throughout the country and in 1923 a Barony was conferred on him, when he took the title Lord Kylsant. The following year, on the death of Lord Pirrie, he became Chairman of Harland & Wolff.

In 1926 the first two motor ships were built in Belfast, the "ASTURIAS" for the Royal Mail Line and the "CARNARVON CASTLE" for Union-Castle. The first appearance of the R.M.M.V. "CARNARVON CASTLE" in Cape Town aroused much interest, with her two squat funnels, a marvellous white panelled First Class Lounge and a completely different lay out to that of the "ARUNDEL

CASTLE". Although the "ASTURIAS" only had one much larger funnel, the engines and propellers were identical and in emergencies spares were often exchanged between the two Companies. Then followed the building of two more motor ships for the Round Africa service, "LLANGIBBY CASTLE" and "DUNBAR CASTLE" who joined the new "LLANDOVERY CASTLE" - her namesake was sunk during the War. On Wednesday 29th June 1927, just before she sailed on her maiden voyage, the "CARNARVON CASTLE" embarked some very distinguished guests for a cruise around the Isle of Wight. A small booklet was issued to the guests describing the new ship. The Commander was Captain H. Strong RD, RNR, 3rd Officer C.G. Gorringe (later Master), 4th Officer H.A. Deller (later Commodore of the Line, Chief Engineer C.E. Manners, 2nd Engineer T. Stewart (later Asst. Supt. Engineer standing by the building of all future ships in Belfast), Purser R.E. Sutton and Chief Steward F.R. Richardson.

There was accommodation for 1,500 passengers and crew, an electrically operated lift, swimming pool, laundry, ship's shop (called an Emporium) and a dark room. On deck, the ship had the latest gear for quickly launching the life boats and two motor launches which were equipped with wireless and search lights.

239 guests names were listed in the booklet of which eighty-eight were titled. There were two Directors on board, apart from Lord Kylsant - Mr. J.W. Clark and the Rt. Hon. Lord Suffield whose wife (the daughter of Lord Kylsant) launched the ship. Among the names well known at that time were:

Lt. Gen. Sir Robert Baden-Powell GCMQ, GCVO
Dame Clara Butt DBE
The Duke of Abercorn KP (later a Director of Union-Castle)
The Cuban and Belgian Ambassadors
Mr. Robertson Fyffe Gibb (later Chairman of Union-Castle)
Miss Rose Macauley
Mr. John Masefield
The Rt. Hon. Lord Shaw of Dunfermline
The Hon. Justice de Villiers
Vice-Admiral A. Walker-Heaneage-Villers
Sir Henry Wood FRCM

It is impossible in a history such as this to continually list the ships in the Fleet, so it is interesting to note that at this particular period the Union-Castle Fleet comprised the following thirty-eight ships:

"CARNARVON CASTLE"	20,063 tons
"ARUNDEL CASTLE"	19,023 tons
"WINDSOR CASTLE"	18,967 tons
"BALMORAL CASTLE"	13,361 tons
"EDINBURGH CASTLE"	13,330 tons
"KENILWORTH CASTLE"	12,975 tons
"ARMADALE CASTLE"	12,973 tons
"WALMER CASTLE"	12,546 tons
"SAXON"	12,385 tons
"LLANSTEPHAN CASTLE"	11,295 tons
"LLANDAFF CASTLE"	10,786 tons
"LLANDOVERY CASTLE"	10,609 tons
"KILDONAN CASTLE"	9,688 tons
"KINFAUNS CASTLE"	9,656 tons
"DURHAM CASTLE"	8,239 tons
"DUNLUCE CASTLE"	8,130 tons
"GLOUCESTER CASTLE"	7,999 tons
"GUILDFORD CASTLE"	7,995 tons
"GARTH CASTLE"	7,715 tons
"SANDGATE CASTLE"	7,634 tons
"GRANTULLY CASTLE"	7,612 tons
"SANDOWN CASTLE"	7,607 tons
"CHEPSTOW CASTLE"	7,494 tons
"RIPLEY CASTLE"	7,446 tons
"GLENGORM CASTLE"	6,763 tons
"BAMPTON CASTLE"	6,696 tons
"BRATTON CASTLE"	6,429 tons
"BANBURY CASTLE"	6,429 tons
"GAIKA"	6,392 tons
"GASCON"	6,341 tons
"CARLOW CASTLE"	5,833 tons

WINCHESTER CASTLE

"DUNDRUM CASTLE"	5,259 tons
"DROMORE CASTLE"	5,242 tons
"CORFE CASTLE"	4,592 tons
"CRAWFORD CASTLE"	4,383 tons
New Steamer building	1,300 tons
"EIDER"	1,236 tons
"HANSA"	880 tons

It is surprising how many Castles there were in the United Kingdom. The last two named and the one being built were 'feeder' ships (known as the black boats because of their black hulls) and they were engaged in taking cargo from the mail vessels in Southampton to Continental ports. It is interesting to note that twenty-seven years after the amalgamation of the Companies, there were still three Union Line ships in the Fleet, sailing under their original names.

A new freight contract was signed in 1928 undertaking to provide improved weekly refrigerated cargo space for perishables such as fruit and meat on the Northbound journey. As a result two more 20,000 ton refrigerated motor-ships were ordered from Belfast, the "WINCHESTER CASTLE" in 1930 and the "WARWICK CASTLE" in 1931. The "WINCHESTER CASTLE" was very similar in design to the "CARNARVON CASTLE", but in the First Class Lounge her panelling was of the Tudor style with heavy drapes and a marvellous marble fireplace in the centre (not that it was ever used as such). With the introduction of these modern liners the "WALMER CASTLE" and "SAXON" were withdrawn from service, the "WALMER CASTLE" being broken up at Blyth in February 1932 and the "SAXON" placed in reserve before finally being scrapped in 1935.

In the meantime Lord Kylsant was still increasing his shipping empire. He had taken over the Bullard King Line, Glen Line, Lamport & Holt, Coast Lines, Australian & Commonwealth Line, Shaw Savill and the Ocean Steam Navigation Company (better known as the White Star Line). He was now Chairman of thirty-seven major shipping Companies owning some 2,000,000 tons of shipping. New Group offices were built in Leadenhall Street although Union-Castle remained in Fenchurch Street.

He now embarked on a large shipbuilding programme with the financial assistance of the Government's Trade Facilities Act involving a loan of millions of pounds (the Act was designed to help the unemployed). However, a depression set in and 4,000,000 tons of shipping was laid up in world ports.

When the repayment of the loan became due the Government proved unwilling to extend the period of payment and, in order to raise money, the public were invited by Lord Kylsant to subscribe to the issue of debenture stock in the parent company - Royal Mail Line. The wording of the prospectus apparently gave the impression that Royal Mail Line was trading with profit. Soon rumours began to circulate in the City of London about the true financial situation of the Group. This resulted in 1931 with a Summons being issued against him.

He was brought to trial at the Central Criminal Court in London under Section 84 of the Larceny Act of 1861 (since repealed) for failing to omit information about his Company's affairs with the result that the prospectus for the issue of debenture stock issued by the company gave a false impression of the Company's position. He was sentenced to twelve month's imprisonment but on other charges, regarding publishing accounts he knew were false, he was acquitted. He appealed against his sentence but the Appeal Judges, meeting in the first week in November 1931, upheld the verdict. On leaving prison the following year, Lord Kylsant returned to his home in South Wales and died in 1937.

In October 1931 Lord Kylsant resigned as Chairman of Union-Castle Line and was replaced, for a short period, by Lord Runciman M.P. until, after a General Election, he joined the Government and was replaced in June 1932 by Mr. Robertson Fyffe Gibb.

Mr. Gibb had started with the Union Line as a boy of fifteen and was eventually placed in charge of the Publicity Department when he produced a House Magazine called *Union-Castle Gazette* (a forerunner of the *Clansman*). A keen sportsman he captained the Company's cricket team and liked to play host to visiting sports teams on the mail ships at Southampton. He later became head of the Passenger Department, but became best known when he was appointed

Manager of the Southampton office. On the takeover by Royal Mail Line he was recalled to London to become Joint Manager of the Company and was later appointed Director.

Mr. Gibb was not an expert in financial matters, but on him, as Chairman, was placed the responsibility of getting the Company back into good standing. No dividends were paid for several years and accordingly a representative of the Shareholders was appointed a Director, Sir F. Vernon Thomson KBE, later himself to become Chairman. He had a background of shipping and finance and, at one time, was Chairman of the King Line.

STIRLING CASTLE

CHAPTER NINE

UNION-CASTLE ON ITS OWN

By November 1934 the effects of the Royal Mail disaster had been overcome and Union-Castle finances were again sound and in July 1936, with a new Board of Directors, the Company was again wholly independent. Although the passenger trade was still affected by the depression the cargo trade was expanding due to the fruit imports from South Africa, so two new refrigerated cargo vessels were ordered - the "ROSLIN CASTLE" and the "ROTHESAY CASTLE". These were the first of many 'R' Class ships which were to last until the end of the Company. At certain periods they were laid up in British ports between the deciduous and citrus fruit seasons and did yeoman work for the Company. The "ROSLIN CASTLE" in particular sailed regularly to and from South Africa (including War service) until she was scrapped in 1968.

In addition two new mail vessels were ordered from Harland & Wolff - the "STIRLING CASTLE" and the "ATHLONE CASTLE" - each approximately 25,000 tons. They each had one raked funnel, two masts, a stream lined superstructure and of such rakish design that they resembled large yachts. They were both diesel motor ships and had much improved passenger accommodation and an innovation was substituting a Cabin Class for the old Tourist Class.

The "STIRLING CASTLE" was launched by Mrs. Robertson Fyffe Gibb, wife of the Chairman, and an innovation was an elegant First Class Drawing Room. When the ship entered the mail service in February 1936 she beat the "SCOT"'s record of 1883 by arriving in Cape Town in thirteen days.

The "ATHLONE CASTLE" was launched by H.R.H. Princess Athlone whose husband, the Earl of Athlone brother of H.M. Queen Mary, had previously been Governor-General of South Africa.

In 1937 another new Mail Contract was signed, this time to last for ten years, when the journey time to the Cape was fixed at not more than fourteen days. In fact, only the "STIRLING CASTLE" and the

DUNNOTTAR CASTLE

"ATHLONE CASTLE" could maintain the necessary speed so another vessel of the same type was ordered and other mail ships were reengined. The two coal-burning steamers, "ARUNDEL CASTLE" and "WINDSOR CASTLE", were sent to Belfast in turn to convert to oil-burning, both had two of their four funnels removed and they were lengthened to give a more rakish bow to allow for more speed. The motor ships "CARNARVON CASTLE", "WINCHESTER CASTLE" and "WARWICK CASTLE" were the next to be sent for re-engining, the first two having their two squat funnels replaced with a single slightly larger one. The "CARNARVON CASTLE" was also lengthened.

In 1934 two new Intermediate ships were also ordered - the "DUNNOTTAR CASTLE" and a new "DUNVEGAN CASTLE". At 15,000 tons they were the largest yet for the Round Africa Service and designed to have a speed equivalent to the older mail ships whom they could replace if necessary. In fact, they were temporarily engaged as such whilst the other ships were being reengined and in doing so took only seventeen days for the outward journey.

In 1937 two even large Intermediaries were ordered with greatly improved passenger facilities which gave the Round Africa Service and East Africa, in particular, its biggest Fleet yet. The 17,832 ton "DURBAN CASTLE" sailed in 1938 followed the following year by the "PRETORIA CASTLE". The latter ship was sold to the Admiralty in 1943 and converted to an aircraft carrier, but re-acquired after the War and after an extensive refit re-entered the Union-Castle Fleet as the "WARWICK CASTLE".

Further tonnage was ordered including four more 'R' Class refrigerated cargo ships - "ROCHESTER CASTLE", "ROXBURGH CASTLE", "RICHMOND CASTLE" and "ROWALLAN CASTLE". Three 'L' Class ships - "LLANDOVERY CASTLE", "LLANSTEPHAN CASTLE" and "LLANDAFF CASTLE" - were all sent to Belfast for conversion from coal-burning to oil-burning. Another new feeder ship for the Southampton-Continent route was ordered, the 900 ton "WALMER CASTLE" to replace the thirty-six year old "EIDER".

CAPETOWN CASTLE and IMPERIAL AIRWAYS FLYING BOAT

The third new mail vessel ordered from Harland & Wolff, Belfast, to complete the fleet was named "CAPETOWN CASTLE" and was the first to be named after a Castle outside of the United Kingdom. The 27,000 ton liner was very similar in design to her two predecessors and was launched on 23rd September 1937 by Mrs. J. Lowe, Mayoress of Cape Town, and she sailed on her maiden voyage on 29th April 1938 under the command of Captain J.A. Barron. Because of her almost perfect lines she was often referred to as the 'Queen of the Southern Seas'. The smallest ship in the Fleet was the 1,289 "ROVUMA" and was based at Beira to bring cargo in from local ports and sometimes to connect with the Mail Service at Durban. She was painted black with one funnel and her moment of glory came in 1939 when, dressed overall, she was in attendance at the arrival in Beira of the President of Portugal. Union-Castle sold her in 1949, but she reappeared later in Cape Town as the "BOUNDARY".

With the completion of the rebuilding and modernisation programme, producing one of the most modern fleets in the world, the sailing time of mail ships was changed to 4.00pm on Thursdays. It was often said by people in Southampton that when the ship's sirens sounded on departure they could set their watches at 4.00pm.

Unfortunately, world events were to change all shipping schedules.

CHAPTER TEN

AT WAR AGAIN

Early in 1939 Mr. Robertson Fyffe Gibb retired as Chairman (although he continued as a Director) and Sir F. Vernon Thomson was elected Chairman. Within a few months, however, saw the commencement of the Second World War and from then until the end of the War he was the Chief Merchant Shipping Adviser, first to the Board of Trade, then called the Ministry of Shipping and later became the Ministry of War Transport, advising the Government on the building, design and need of all merchant shipping. His personal assistant was Mr. John Bevan of the Union-Castle Line, later to become General Manager. In 1939, Sir Vernon Thomson was also President of the Chamber of Shipping. Before the War was actually declared, the Admiralty requisitioned the 13,000 ton "EDINBURGH CASTLE" which had been in reserve at Netley for conversion to an Armed Merchant Cruiser. Also requisitioned was the Southampton launch "FALCON". This was followed shortly afterwards by the "DUNNOTTAR CASTLE" which was actually loading in London for a voyage to East Africa and the Cape. The cargo was discharged and she was sent to Belfast, also for conversion to an Armed Merchant Cruiser. To accommodate her passengers and cargo the 8,000 ton "GLOUCESTER CASTLE" was brought out of reserve to sail once more to the Cape.

On the 3rd September the "DUNVEGAN CASTLE" was at East London ready to sail to Durban and the UK, via the Suez Canal. Passengers and a cargo were all on board and the gangway about to be lowered when a cable arrived for the Captain ordering him to return directly to the UK. Passengers and baggage were put ashore and she sailed direct to Table Bay, where her East African cargo was discharged and she shipped more essential cargo for Britain. Both she and the "PRETORIA CASTLE" at the next berth were painted in wartime grey and, on arrival, were sent to Belfast for conversion as Armed Merchant Cruisers.

The next ship affected, two weeks later, was the "CARNARVON CASTLE" when, on arrival at Cape Town, her passengers were told they could go no further. They were disembarked and the cargo discharged. All furniture from public rooms and cabins was off-loaded and she was handed over to the Royal Navy at Simonstown, where she was converted to a fighting ship with eight six-inch guns and other armaments.

For a while after the Declaration of Hostilities, passengers and mail services to South Africa continued with only minor delays being experienced. However, this was not for long and in January 1940 the first Union-Castle vessel was sunk. She was the "DUNBAR CASTLE" and, whilst en route from London to the Cape, she was sunk by a magnetic mine off Deal, in Kent, and for many years afterwards there were still signs of her mast-heads in the water.

Another ship to play an active part in the War was the 5,200 ton cargo ship "DUNDRUM CASTLE". In June 1940 she was in St. Nazaire with a military cargo for Dunkirk (although by then that port was no longer in use). She was about to sail when 500 troops were put on board. Somewhat delayed, she was heading for the harbour entrance when German bombers attacked the port concentrating on the shipping. The White Star liner "LANCASTRIA", with 5,000 troops on board being evacuated from France, received a direct hit, turned on her side and sank. The "DUNDRUM CASTLE" made for the scene and lowered two lifeboats to help pick up survivors, which proved to be slow and hazardous work since the water was thick with oil. Altogether she picked up 120 survivors before joining a night convoy to Southampton.

By 1940 all British ships were in Government service of some kind and in the December the then "H.M.S. CARNARVON CASTLE" distinguished herself with an encounter with the German raider "THOR" in the South Atlantic. These raiders operated alone, far from their base and in unexpected waters and the "THOR" had already claimed several victims.

Both ships fired broadsides and were both hit, on "CARNARVON CASTLE" the electric control gear was shot away and the guns had to be fired independently by hand. "THOR" fired two torpedoes which luckily missed their target and the former liner was scoring repeated hits on the raider which, after ninety minutes of action, escaped through a smoke-screen. The "CARNARVON CASTLE" had been hit thirty-eight times, six crew were killed and thirty-two wounded. She was given permission to stay in Montevideo for forty-eight hours for repairs and it was with some satisfaction to the crew that the holes in the ship's sides were patched up with plates from the former German battleship "ADMIRAL GRAF SPEE", which had been scuttled in the harbour the previous year. From there the liner returned to Cape Town for permanent repairs before resuming her Naval duties.

It had been in the later stages of the First War that German submarines proved to be a menace, but now they were a threat from the beginning, hunting as they were in packs together with surface raiders, magnetic mines and dive-bombers. Every voyage for the merchant ships was now fraught with danger. Slower cargo ships usually sailed in convoys with a Naval escort, but it was often considered that the former mail ships had sufficient speed to sail alone. There was no dazzle paint now and all ships were painted battleship grey. The Admiralty soon began to realise that Armed Merchant Cruisers were not the answer to the attacks. In August 1940 the "DUNVEGAN CASTLE" had been sunk off Northern Ireland and the "DUNNOTTAR CASTLE" was handed back for trooping duties. The Admiralty purchased the "PRETORIA CASTLE" and converted her to an auxiliary aircraft carrier which, of course, completely altered her outline.

As "H.M.S. PRETORIA CASTLE" she was not a conventional aircraft carrier since, under the Command of Vice-Admiral R.B. Davies VC, CB, DSO, AFC, RN, she acted as an experimental carrier. Every new type of aircraft for the Fleet Air Arm, together with other ancillary equipment (armour plating, guns, life jackets etc), underwent exacting tests on board before being used operationally. Representatives of British and American manufacturers were also on board to watch their products under test. The Commander of Flying, Commander P. Bramwell, DSO, DSC, RN, was one of the Royal Navy's most experienced pilots.

ATHLONE CASTLE

The West African port of Sierra Leone was at the centre of the Cape convoy route and they often broke their journey there. To accommodate the increased Naval Staff at the port, the "EDINBURGH CASTLE" was sent there as a Depot ship where she remained throughout the War. In September 1945, her Naval service over and her passenger carrying completed many years previously, she was ceremoniously sunk off the port by British gunfire.

Many other liners were adapted to serve as troop ships with bunks fitted into holds, swimming pools and public rooms. The "STIRLING CASTLE", for instance, had been designed to accommodate 700 passengers, but now carried over 5,000 troops, but obviously not to the same standards.

To supply the Army in North Africa with troops and equipment, convoys left England and sailed the long way round, via the Cape. In one convoy, six Union-Castle ships were included with the "ATHLONE CASTLE" acting as Commodore ship. The usual practice en route was for half the convoy to call in at Cape Town and the other half at Durban, at both ports the crews and troops enjoyed the hospitality of the local people.

In June 1941 Germany attacked Russia and a new phase of the War began with a new expression which brought dread in the hearts of many seamen, both Royal and Merchant - "Russian convoy" - with the extreme dangers during the voyage and the ice and cold on arrival. In the first of these convoys, the "LLANSTEPHAN CASTLE" was the Commodore ship having on board the Commodore and 500 men of the Royal Air Force. On leaving Scapa Flow it was revealed that the convoy was heading to Iceland and Archangel. Being the first convoy it proved uneventful, but with subsequent convoys the German Navy went all out to prevent their arrival in Russia. The "LLANSTEPHAN CASTLE" stayed there for a month while aircraft sent out from Britain were assembled and flown to Murmansk. She returned home with 200 Polish airmen on board who had been rescued from Russian camps and were now to join the R.A.F.

Throughout the War, the "LLANDOVERY CASTLE" served as a hospital ship in various theatres of operations. In May 1942 she brought to Alexandria British prisoners of war who had been exchanged for Italians captured in North Africa.

At the end of 1942 came the next offensive - the Allied landings in North Africa, in which several Union-Castle ships took part. The "WINCHESTER CASTLE" with British and American Commandos on board, together with "WARWICK CASTLE" headed for Algiers. The "DURBAN CASTLE" was detailed as headquarters ships and together with "LLANGIBBY CASTLE" led a convoy into Arzeu Bay. Earlier, as they had all entered the Mediterranean, the combined vessels of the invasion force stretched as far as the eye could see.

Despite heavy air attacks on the ships, the landings were successfully completed and the troop ships returned to Gibraltar to join a homeward bound convoy, but en route the "WARWICK CASTLE" was torpedoed.

The weather was extremely bad and it took up to six hours to rescue the survivors. The Commander, Captain H.L. Shaw, was among several picked up from a raft, but he died shortly afterwards. One landing craft was adrift for six days without food and water before being rescued.

More troops were embarked in the UK and another convoy sailed to North Africa. Included was the "WINDSOR CASTLE" and the convoy was sailing north of Algiers when she was hit by a torpedo from a German aircraft. The crew and the 2,000 troops on board methodically abandoned ship and were rescued. It had been hoped that a destroyer could take her in tow, but it arrived too late before she slid, stern first, into the water. "DURBAN CASTLE", "LLANGIBBY CASTLE" and "WINCHESTER CASTLE" also took part in the Allied landings in Sicily, Salerno and Anzio.

At the end of 1942 the "CAPETOWN CASTLE" sailed from Bombay to Suez, where she embarked a load of German prisoners of war. She reached the Cape safely, dodging the packs of 'U' boats which patrolled the South African coast, and was sent to Saldanha Bay. From there she made a dash across the South Atlantic to the Magellan Straits into the safer waters of the Pacific Ocean. The 27,000 ton liner then sailed through the Panama Canal and up to New York before crossing to Great Britain.

By 1943 American troops were pouring across the Atlantic and helping to cope with the influx were "CAPETOWN CASTLE", "STIRLING CASTLE" and "ATHLONE CASTLE" who were later joined by "CARNARVON CASTLE" when she had completed her Naval service. On one voyage "STIRLING CASTLE" embarked 6,160 American troops.

Captain E.H. Thornton RNR, Master of "CAPETOWN CASTLE" and Commodore of the Line, received a Knighthood in the King's Birthday Honours in June 1944. In the same list Captain F.A.G. Hunter, former Cargo Superintendent at Southampton, was awarded the DSC.

CHAPTER ELEVEN

SOME EPIC WAR-TIME VOYAGES

One of the most famous voyages of a Union-Castle ship was that of the "LLANGIBBY CASTLE" in 1942. Following Pearl Harbour and the entry of America in the War, the Far East became an important theatre of operations, and a large convoy left Britain with troops bound for Singapore. Four days after leaving port the "LLANGIBBY CASTLE" was hit in the stern by a torpedo resulting in the death of twenty-six men with four wounded, her stern defensive gun destroyed and even more serious was the fact that the rudder had been hit and the only means of steering was by means of the engines.

There were 1,500 troops on board and a strong wind was blowing with high seas. No escort was available so she was on her own in 'U' boat infected waters. Captain R.F. Bayer decided to make for the Azores, 700 miles away, at a speed of 9 knots. Within hours a German bomber attacked the ship, but the bombs missed and it was believed the plane was hit by the AA defences on board.

It took three days to reach the islands where they were allowed to stay for fourteen days for temporary repairs. She left on 2nd February with three destroyers and an Admiralty tug as escort, but almost immediately the destroyers were called away to deal with local 'U' boat activity. The "LLANGIBBY CASTLE" developed engine trouble and the tug took her in tow, but the next morning she released the tow and again steered herself with her engines, making very erratic progress which may have made her safer from attack. On 6th February she received an air escort and, two days later, docked safely at Gibraltar where her service personnel were disembarked. During her fifty-seven day stay there more repairs were made, but there was no possibility of a new rudder so she left under her own steam for the last 1,445 miles to the UK. The journey took six days and altogether she steamed 3,400 miles without a rudder.

Another mail ship to have an adventurous war-time career was the "WINCHESTER CASTLE". After a long trooping voyage to

Bombay she returned to Glasgow in May 1940 and commenced to load battle equipment with, mysteriously, a Royal Navy Officer and several Royal Marines in attendance to ensure the correct order of stowage. The crew began to realise that something different was afoot when she sailed to Loch Fyne, where her eighteen life boats were landed and replaced with landing craft before proceeding to Inverary where the HQ of Combined Operations had been established. The "WINCHESTER CASTLE" stayed in these waters for a year carrying out various exercises and manoeuvres and was visited by Lord Louis Mountbatten and Admiral Sir Roger Keyes.

She returned to sea in March 1942 with commandos and other specialist troops bound for Madagascar in the Indian Ocean in a very heavily guarded convoy. At Durban she was joined by the "LLAND-AFF CASTLE" with 1,150 troops on board (she had recently undergone a refit at Durban when her Tourist Class accommodation and most of the First Class had been removed and three troop decks installed). Madagascar has pro-German leanings and, with its fine port of Diego Suarez, it was regarded as a potential submarine base for enemy submarines, especially the Japanese.

At the end of April 1942 the convoy sailed under heavy escort with "WINCHESTER CASTLE" as headquarters ship and, on arrival, the troops were successfully landed together with some engineers from the ship who helped in saving and servicing captured vessels. Within a short time the island was made secure.

The "WINCHESTER CASTLE" stayed in the island before eventually she sailed to New York for a refit before returning to Lock Fyne to prepare for even greater things to come.

After her Madagascar adventure the "LLANDAFF CASTLE" returned to her former duties of carrying troops from the Cape to Suez. In November 1942 she returned from Suez with troops and called in at Mombasa and Dar-es-Salaam before heading for Durban, but whilst off the coast of Zanzibar was sunk by submarine. The crew, under Captain J.C. Clutterbuck, and troops made an orderly evacuation in the life boats and rafts. By daylight they had formed themselves into a convoy and after three and a half hours were rescued by destroyers. Two lifeboats were unaccounted for, one later being

ROCHESTER CASTLE

picked up by another destroyer. The second had decided to head for the shore and the spot where they landed was very wild and rugged and it appeared deserted. Luckily, two Zulu boys working in a field saw them land and returned to their native village. Their father, Mpahleni Zikali, sent food down to the beach for the survivors and at the same time sent runners to the nearest town for assistance.

In due course some sort of motorised transport made its way across country to pick them up and take them to the railway station at Umkuzi where they caught a train to Durban.

After the War the Administrator of the Province heard that Zikali had a grievance since his part in the rescue had not been recognised. He gave back a compass from the "LLANDAFF CASTLE"'s life boat saying that he did not want a watch that did not keep the proper time. The Union-Castle office at Durban was contacted, they then reported it to London and Sir Vernon Thomson gave instructions that Zikali be properly rewarded. Subsequently, he received three live cattle and a few other gifts as well as a roasted ox to provide a feast for all those in the village who had helped rescue the survivors.

One of the most heroic and hazardous voyages made by a Union-Castle ship was not by a liner but by one of the 'R' Class cargo ships - the "ROCHESTER CASTLE". By 1941 the island of Malta was a thorn in the side of the Germans and Italians trying to win the battle in North Africa so the island was almost bombed out of existence. The dockyards were badly damaged and little food was available or fuel for the Spitfires trying to defend the island. The only relief came from sea convoys whose passage became very perilous. Early in 1942 eighty-five merchant ships sailed from England, of which, twenty-four were sunk and eleven returned to port.

In the August of that year a convoy of fourteen merchantmen (including "ROCHESTER CASTLE") sailed from the Clyde on Operation Pedestal, the crews having been told that if Malta and the North African campaign were to survive, the convoy must get through. Escorts were the battleships "RODNEY" and "NELSON", cruisers "NIGERIA" and "MANCHESTER" together with twelve destroyers. Later the convoy was joined by the aircraft carriers

"INDOMINITABLE", "EAGLE" and "VICTORIOUS", five more cruisers and twenty more destroyers.

After a week at sea, they passed through the Straits of Gibraltar without incident, but two days later, on 11th August, enemy aircraft were spotted shadowing the convoy. That afternoon the aircraft carrier "EAGLE" was sunk. "ROCHESTER CASTLE" was armed with a Bofors gun, a 4.7 inch low angle gun, four machine guns, four FAMS (rocket-fired parachutes on a length of wire with an explosive attached) and a 3 inch rocket gun. Another Bofors gun, part of the cargo, had also been mounted on No. 2 hatch. Air attacks were now continuous and the "ROCHESTER CASTLE" was straddled with bombs and water was flooding into the engine-room. There was no relief for the gun crews who were given makeshift meals at their guns.

On 12th August came more bombers dropping mines as well as bombs and a submarine sank the first merchant ship, "CLAN FERGUSON" (although all her crew were rescued) as well as a cruiser and an AA cruiser. That evening the "ROCHESTER CASTLE" was hit by a torpedo fired by an 'E' boat which was sunk by gunfire from the ship before it could escape. One of her holds was flooded, but she still managed to make 13½ knots and at 6.00pm the following day she led the battered survivors into Valetta Harbour to be welcomed by some 70,000 people lining the harbour. In her side was a hole 20ft by 18ft. Her Master - Captain R. Wren, Chief Officer - Mr. A.J. Culpin, Chief Engineer - Mr. F.E. Mattocks and Chief Steward - Mr. Harry White were all decorated for their part in the success of the voyage. Altogether nine merchant ships were lost together with several Royal Navy vessels.

When the ship was finally scrapped, her ship's bell was presented to the Merchant Navy Seafarers Memorial Club in Malta where it has been proudly displayed ever since. Some years later the Chief Steward, Harry White, came ashore to be Housekeeper at the Passenger Department's office at Rotherwick House in London and when he retired paid a visit to Malta and was very pleased to see the bell again.

Five Company ships were lost in 1942, "RICHMOND CASTLE", "ROWALLAN CASTLE", "ROXBURGH CASTLE" and

"LLANDAFF CASTLE". The "ROXBURGH CASTLE" was outward bound to Rio de Janeiro when on the fourth day of the voyage she was torpedoed. The crew took to the boats and half an hour later the ship sank. The 'U' boat surfaced and her Commander apologised for sinking them before dispensing food and water and the directions to the Azores before, once again, diving.

Another ship lost was the "GLOUCESTER CASTLE" which mysteriously disappeared. She had been taken out of reserve and put back into service to the Cape. However, she had no passenger certificate and could only carry twelve passengers. She sailed in June 1942 from Birkenhead direct to Cape Town, Port Elizabeth and Durban. She sailed in convoy for some days but, because of her slowness, she proceeded unescorted to the Cape but she never reached her destination. Two months later the relatives of the passengers and crew were told that the ship was now seriously overdue and all must be presumed dead.

It was not until after the War that the true story became known, but that proved a dilemma for the Superannuation Fund as to whether or not to pay the possible widows a pension.

On 15th July 1942 a still and overcast night off Ascension Island, a Germany raider came in close and shelled the liner. The first shell wrecked the Bridge and Radio Room and killed most of the Deck Officers, the second hit the dining saloon and ignited petrol stowed on deck. She sank within minutes and of the 154 on board, ninety-three had died including eight passengers. One woman and three children only survived. The raider sent off a motor boat for survivors and the one life boat which managed to get away also made for the raider as the only means of rescue. On board the raider were also survivors from other ships and later they were all transferred to a German tanker acting as a reception centre for prisoners captured by a group of raiders in that area.

Conditions on board were horrific and food almost unfit to eat, with sanitation at its most basic. The tanker was stationary for a month before another group of prisoners arrived from a British ship called the "GEMSTONE", including the Master, a Captain Griffiths, who was later allowed to return to the raider which had picked him up. The tanker then sailed for Japan but, on calling at Singapore en route, the Japanese there took over forty prisoners to work ashore including most of the surviving crew of the "GLOUCESTER CASTLE". The tanker eventually reached Japan where the remaining prisoners were interned.

In Singapore, the crew of the ship suffered severe hardship and were forced to do very rigorous work with little food. Captain Griffiths however, was eventually interned in Germany and managed to write to his wife giving the names of some of the people he had briefly met on the tanker. Union-Castle Line were informed that some members of the crew had survived including Captain Rose and the Red Cross contacted their relatives.

Plans were now going ahead for the greatest assault of all - the invasion of Normandy. Amongst other ships involved, the "LLANGIBBY CASTLE" returned to the Clyde for a thorough overhaul including the shipping of landing craft and being painted an 'early morning' camouflage. She sailed to Milford Haven to embark further specialised troops in addition to the specially trained body of Canadian troops and a detachment of Royal Marines already on board and from there to the Solent. Southampton Water and Spithead was the scene of intense activity and the "LLANGIBBY CASTLE", which was attached to a unit called 'J' Force, spent much time in exercises, night manoeuvres and signalling, ending with a 'dummy' invasion of Bracklesham Bay.

Captain T.W. McAllen attended the final briefing at the HQ at Cowes when final details were given and 'D' Day fixed for Monday 5th June. However, bad weather caused its postponement for twenty-four hours and on the evening of 5th June 'J' Force sailed with "LLANGIBBY CASTLE" and arrived off the beach-head in Normandy at 5.30am the next morning where her troops were all successfully landed. Since she had the largest number of troops on board it was 3.00pm before she left the Normandy coast and arrived back in Spithead at 9.00am the following day to pick up more troops. Altogether she crossed the Channel more than sixty times carrying troops to France estimated at more than 100,000.

The "LLANDOVERY CASTLE", still a hospital ship, had been brought to Southampton and sailed with the convoys to attend the wounded. After the port of Cherbourg was free she embarked there many of the wounded Canadians to return them direct to Montreal. Larger troop ships also arrived in Cherbourg, including the "DUNNOTTAR CASTLE", and they could land the troops without them getting their feet wet.

Among other ships to use Cherbourg was the "ARUNDEL CASTLE" and on 15th September 1944 she arrived at Liverpool under Red Cross markings, with repatriated prisoners of war.

Southampton was undoubtedly the centre of activities for 'D' Day with landing and assault craft berthed in every available space. It was said at the time that one could almost walk from Southampton Docks to Marchwood by stepping over these craft. Every inlet on the South Coast as far as Christchurch Bay was also full of shipping.

The Ministry of War Transport had asked the Union-Castle Line to set up a stevedoring organisation for the loading and unloading of equipment. The Cargo Department returned from Liverpool and Glasgow and, in all, they handled two million tons of equipment and for lashings used 2,284 miles of wire.

In the early days of the War air attacks were expected in London and some of the Departments in the London Office and many of the records and archives were transferred to King's Court, a wonderful stately building in its own grounds at Chandlers Ford, just outside Southampton. Most of the Southampton staff who had not been called up were transferred to Liverpool, Glasgow and King's Court. The Staff and Property Departments were moved from London to temporary premises in Watford.

In Cape Town with so many different ships arriving during the War, the Union-Castle Marine Superintendents at the various ports were given the task of setting up and organising a Ministry of War Transport in South Africa. The Company also acted as agents for many of the shipping companies not normally represented there.

Throughout the hostilities the Company lost eleven ships through enemy action, six of them passenger liners. Also lost was the

900 ton "WALMER CASTLE", the last of the feeder service to the Continent which was not continued after the War.

Apart from the activities of the Armed Merchant Cruisers and other ships under the direction of the Royal Navy, Union-Castle ships sailed nearly seven million miles during the War - the "ATHLONE CASTLE" alone travelled half a million miles. Between them they carried 1,290,000 troops and civilians. Of all the ships involved, it would seem that the "LLANGIBBY CASTLE" was in more action than others.

Of the Company's staff called to the Colours, sixty-five died - 209 listed as missing, thirty-five wounded and sixty-two taken prisoner. No fewer than 144 received honours and awards and a further fifty-nine mentioned in despatches.

CHAPTER TWELVE

THE IMMEDIATE POST-WAR YEARS

After VE day in Europe the "LLANSTEPHAN CASTLE", which had been taken over by the Royal Indian Navy, played her part in the campaign in Burma. When the War was finally over the Company's ships were still employed carrying troops all over the world, before they could resume their normal duties.

By this time South Africa had an urgent need for skilled technicians and sponsored a scheme whereby engineers and skilled workers, with their families, could emigrate from the UK. Accordingly the "CARNARVON CASTLE", "WINCHESTER CASTLE" and "ARUNDEL CASTLE" were employed on this service and carried 32,000 emigrants to South Africa, not perhaps to pre-War standards, but certainly better than in their trooping days.

It was 1947 before passenger services were fully restored with the sailing of the "CAPETOWN CASTLE" on the mail run and the "LLANDOVERY CASTLE" on the Round Africa trip and the following year before "ARUNDEL CASTLE" returned from conversion from her War time duties. In the meantime, several of the 'R' Class cargo ships carried the mail.

In the final stages of the War, in 1945, Sir Vernon Thomson flew to South Africa to negotiate new mail and freight contracts, both of which became operative on the 1st January 1947 to last ten years, the length of the journey to take no more than fourteen days. The Company also agreed to have certain of their cargo ships registered in the Cape and to employ more South African nationals on their ships.

To do this Union-Castle initiated a scheme for carrying cadets, including some of those who had excelled in the South African training ship "GENERAL BOTHA". One of these cadets was Norman M. Lloyd, who later became Master of the "S.A. VAAL" and was appointed Commodore of the Safmarine Fleet.

Despite the fact that the post-War years produced an increase in air traffic to South Africa also carrying mail, the mail carried by the

EDINBURGH CASTLE

Company actually increased even though it was now only second-class including newspapers and parcels.

Another re-building programme was obviously necessary which resulted in thirteen new Castles - "PRETORIA" and "EDINBURGH" (Mail), "KENYA", "BRAEMAR" and "RHODESIA" (Intermediate), "BLOEMFONTEIN" (Round Africa), "RIEBEECK" and "RUSTENBURG" (Refrigerated Cargo) and "TANTALLON", "TINTAGEL", "GOOD HOPE", "DRAKENSBERG" and "KENILWORTH" (General Cargo). The last three were purchased from the Government and the "GOOD HOPE CASTLE" and "DRAKENSBERG CASTLE" were registered in Cape Town and flew the South African flag as agreed in the Mail Contract. They were officially handed over at ceremonies in Cape Town in July 1946.

The new mail ships were to replace the "WARWICK CASTLE" and "WINDSOR CASTLE" which were sunk in the War and both were ordered from Harland & Wolff in Belfast. The first was named "PRETORIA CASTLE" and the launching on 19th August 1947 was an innovation for the Company and shipping in general. Usually the naming and launching ceremony is performed by a Lady who throws a bottle of wine against the bows of the ship which then slides into the water. On this occasion the ceremony was performed by Mrs. J.C. Smuts, wife of Field Marshall Jan Smuts - Prime Minister of South Africa, from her home 6,000 miles away in the Transvaal. The naming was performed by radio and after speaking into a microphone, first in English and then in Afrikaans, Mrs. Smuts pressed a button creating electrical impulses to release the bottle of wine to launch the ship. She was represented at the actual ceremony by her daughter, Mrs. Bancroft-Clark.

However, by the time the ship arrived in Cape Town on her first voyage, there had been a General Election and Field Marshall Smuts had been replaced by Dr. D.F. Malan. Nevertheless, Dr. Malan and his wife and Field Marshall Smuts and his wife were all present at a reception on board hosted by Sir Vernon Thomson.

Her sister ship, "EDINBURGH CASTLE", was launched on 16th October 1947 by H.R.H. Princess Margaret, on her first official engagement, and sailed on her maiden voyage on 9th December 1948.

BLOEMFONTEIN CASTLE

Both ships of nearly 28,000 tons were the largest and most powerful in the fleet. They were more or less identical, up graded from "CAPETOWN CASTLE" and reverting once more to oil-fired steam turbine engines with accommodation for 227 First Class passengers, 487 Tourist Class and with a crew of 400.

The next ship to be built was the 19,400 ton "BLOEMFONTEIN CASTLE" for the Round Africa service. She was designed to carry 700 passengers and for the first time in the Company's history she was a one-class ship with passengers having access to all facilities. Many of the junior members of the staff were invited by the Directors to travel to Belfast and sail in her to London prior to her first voyage.

Then followed the building, also in Belfast, of the three new Intermediate vessels (they sailed to South Africa either via the East Coast or the West Coast and returned the same way). The "KENYA CASTLE", "BRAEMAR CASTLE" and "RHODESIA CASTLE" were also one-class ships with accommodation for 530 passengers and, in many ways, their public rooms were smaller versions of "PRETORIA CASTLE" and "EDINBURGH CASTLE". They replaced the three 'L' Class ships then still running from London. The "KENYA CASTLE" undertook a short cruise from Southampton before sailing from London in March 1952.

The two new 'R' Class refrigerated cargo ships, ordered in 1946, were the 8,322 ton "RUSTENBURG CASTLE" and "RIEBECK CASTLE" - the largest yet in that class, with accommodation for twelve passengers. This was followed a few years later by the first two general cargo ships the Company had had built for some years, the 7,300 ton "TINTAGEL CASTLE" and "TANTALLON CASTLE". To replace the thirty year old "ARUNDEL CASTLE", an order was placed on Harland & Wolff, Belfast for an even larger mail liner to be called "PENDENNIS CASTLE", after the ancient Castle in Cornwall.

Back in 1919 the then Chairman, Sir Owen Philipps, purchased Bullard, King & Company Limited, also known as the Natal Line, and which had been founded in the 1850's by a Captain Samuel Bullard and Captain Daniel King. Their ships sailed direct from England to

Durban where they were always able to cross the bar. In more recent years, they extended their journeys to south and East Africa in general.

The Directors of Union-Castle now felt that additional general cargo tonnage was necessary and they acquired a controlling interest in the King Line, followed a few months later by complete ownership. The firm had been founded as the King Alfred Steam Ship Company, but shortly afterwards resorted to the shorter name of King Line. At the time of the purchase they had eleven ships with a dead-weight capacity of 103,000 tons, but in later years they concentrated more on bulk carriers.

There were changes too in the Company's affairs at Southampton. For many years the inward vessel docked at Berth 35 at the far end of the Old Docks and when the cargo had been discharged moved round the 'knuckle' to Berth 38, the outward sailing berth. In between was the Company's Dock Office housing Marine, Engineering, Catering and Stores Department. The main Southampton Office was still in Canute Road. Even now, years after the Company's service finished, it is still called Union-Castle House mainly housing Shipping Agents. In Georgian times the site was occupied by the Gloucester Baths, but in 1847 they were demolished and Southampton's first Custom House was built on the site. The Customs Office then moved to larger premises nearby and the building was taken over by the Union Line in 1892. On the merger, in 1900, it became the Southampton Headquarters of the Union-Castle Line (the Castle Line did not have an office in Southampton and their ships were handled by Agents).

With the increased passenger and cargo carrying capacities of the new mail ships the berths in the Old Docks proved inadequate and with the co-operation of the Docks Board, the Company moved to larger berths in the New Docks with inward ships berthing at 102 Berth and the outward ships sailing from Berth 103, with a luxurious passenger terminal in between. It was designed for passengers waiting to board or, on arrival, their baggage to come from the holds.

Passengers entered or left the very comfortable mahogany-panelled Reception Hall on the first floor via a covered gangway from the ship. Other facilities included an Immigration Room, a circular bank of telephones, a writing room, control room, bar and buffet, bookstall and Union-Castle's stand with an island counter for Travel Agents and railway ticket office. Among the use of the public address system was the calling in to the Terminal of taxis and private cars from the nearby car park.

At the same time a new office complex was built in Herbert Walker Avenue, alongside the berths, complete with restaurant facilities for the staff. When it was completed, under the military style direction of Major-General Prior-Palmer - Area Manager, the whole of the two offices were moved under the same roof, department by department, in just two days. For availability to the general public, the Passenger Department remained in Canute Road.

CHAPTER THIRTEEN

CATERING, PASSENGER AND CREW REGULATIONS

The service provided by Union-Castle Line and their predecessors has always been to a high standard, despite the rather primitive conditions at sea in the early days. The food was usually excellent with plenty of choice, especially in the First Class (there are very few surviving Tourist Class menus).

One of the earliest available menus is dated 26th June 1897 when Queen Victoria's Diamond Jubilee was celebrated by a Royal Navy Review at Spithead. Castle Line's "ARUNDEL CASTLE" was present and, at a reception on board, the following choices were available:

Asparagus soup
Turbot with lobster sauce
Lamb cutlets
Stewed duckling
Roast saddle of lamb
Roast beef
York ham
Braised chicken and Malay sauce

This was followed by:

Victoria pudding
Macedoine of fruit
French pastry
Cheese
Dessert
Coffee

On 8th August 1905 another Royal Navy Review was held, this time by King Edward VII, and included an official visit from the

French Fleet. The Union-Castle liner "ARMADALE CASTLE" acted as hosts to the French Navy and a breakfast menu lists the following:

Quaker Oats
Fish
Grilled kidneys
Bacon
Devilled chicken
Eggs
Tomatoes
Oriental curry
Cold braised beef
York ham

Fruit

Tea
Coffee
Cocoa

A First Class dinner menu of "GLENGORM CASTLE" (Captain E.S. Vincent) dated 15th February 1929 included:

Hor's D'oeuvres

Soup
Halibut
Lambs tongues
Sirloin of beef
Braised pheasant

Pouding Sans Souci
Apricot melba
Paris roll

Curried prawns

An interesting note at the bottom of the menu stated that smoking was not permitted in the Dining Saloon. As often happened then, as now, this particular copy must have been passed round the table and was signed by fifteen of the passengers.

Another menu of "ARMADALE CASTLE" in 1933 comprised the Chef's Menu and an A La Carte menu, the latter comprising a choice of twenty-four items.

An interesting insight into catering and working conditions on board the earlier mail ships recently came to light with the production of two books kept by the late Mr. G.H. Inkson, a former Chief Steward, covering the 1920's. One book, handwritten, shows a complete summary of menus and dishes available together with working instructions. It would appear that there were twenty-nine egg dishes, twenty-five different salads and ninety-four puddings and sweets. A menu card attached was that of the "BALMORAL CASTLE" on 21st July 1919, the occasion being a Parliamentary visit to the Fleet in the Thames. Starting with salmon and dressed crab it proceeded through sirloin beef, hindquarter of lamb, galantine of capon, roast chicken, York ham and ox-tongue to greengage tart, ice pudding, cheese and fruit with coffee being served on deck. Wines were Chablis, Le Pavillon Chateau Margaux 1914, Pommard 1914 and liqueurs.

On another smaller card addressed to the Commander, Captain Sir Benjamin Chave KBE, the menu was prefixed with the following verse:

You may sup with the Gods on Olympus
And quaff their ambrosial wines
But tonight in the seas
Your good hosts do agree
That from the graceful springbok
And gold juice from a wine
Your portion must be....

This was followed by a six course menu, the main item being springbok.

At the other end of the scale, a typical day's food in the Third Class Saloon was:

Breakfast:	Kippers Sausages and mash American hash cakes
Lunch:	Rice Mutton and potato pie Breast of lamb
Tea:	Soused herrings Roast sirloin of beef Leg of pork Pickles Scones

Union Day menus on board "SAXON" in 1925 and "WALMER CASTLE" in 1927 were printed in English and Afrikaans. In more modern times menus became more sophisticated and were sometimes printed in French with the English underneath. Up to the 1960's, when many of our Mail and Intermediate ships had a Minstrel's gallery above the First Class Dining Saloon, the menu would often include the musical programme to be played during Dinner.

In the latter years of the service, the mail ships produced a Farewell Menu for the last night of the voyage which, apart from a luxurious menu, also included an extract from the Ship's Log giving weather and navigational details. For these, and other important events, the menu covers usually depicted a picture of the ship. The covers of other menus were usually of South African birds, flowers and wildlife.

One of the earliest of these Farewell Menus was that of the "EDINBURGH CASTLE" with A. Harrop as Chief Catering Officer and H. Laight as Chef de Cuisine.

To end a short history of shipboard menus, it is appropriate to re-produce the last one of all. The final voyage of "S.A. VAAL" on Sunday 9th October 1977.

See attached menu

"S. A. VAAL" (30,212 GROSS TONS)

"FAREWELL DINNER"

To Mark the Termination
of the
Passenger Mailship Service

1857 - 1977

Safmarine

1857 *Diner* 1977

Saumon Fume au Crevettes Melon Frappe

Consomme aux Xeres Creme de Tamate

Scampi Provencale

Supreme de Sole, Savoy

Cotelettes d'Agneau, Reforme

Contre-filet de Boeuf, Sauce Raifort

Dinde Roti, Anglaise

Tournedos, Rossini

Pommes - Fondante et Persillees

Choux de Bruxelles Courgettes Fines Herbes

Desserts

Pouding Anglaise, Sauce Brandy

Salade de Fruits au Liqueur

Profiterolles au Chocolat

Meringue Glace Fraises

Glace

Vanille Cafe d'Ananas

Sauce Chocolat

Savoureux

Champignon sur Croute

Fromage

Fruite

Cafe

SUNDAY OCTOBER 9th 1977

Extracts from the Log of the Final Northbound Voyage of the "S.A. Vaal"

"Voyage Number, One Hundred and Fourteen"

LAUNCHED CLYDEBANK 17-1-61 COMMODORE R. W. McA. THOMSON

CAPE TOWN TO SOUTHAMPTON

Date	Day's Run	Lat. / Long.	Wind Force	Temp. at Noon Air	Sea	Itinerary
27-9-77	At Cape Town		Var 3	21.1°	—	1600 Departed Cape Town 2021 Passed Cape Columbine Cloudless
28-9-77	423	28°19S 13°22E	SE 6	18.0°	15.6°	Ship's Clocks retarded one hour to GMT plus one hour Commodore's Cocktail Party. Welcome Aboard Dance Sunny with few clouds Low S'ly swell
29-9-77	509	21°24S 07°57E	SWxW 2/3	17.1°	16.7°	0506 Crossed Tropic of Capricorn and entered Tropics Cloudy and clear Low S'ly swell Frog Racing
30-9-77	513	14°36S 02°30E	SE 3	18.6°	18.3°	Tropical Night on the Promenade Deck 2230 Crossed Greenwich Meridian and entered Western Hemisphere. Overcast but clearing Low SxE'ly swell
1-10-77	508	08°03S 2°59W	SExS 3/4	22.1°	23.3°	Vaal Revue Sunny with cloudy patches Low SSE'ly swell
2-10-77	508	01°31S 08°24W	SSE 3/4	24.2°	23.3°	1030 Divine Service conducted by the Commodore 1740 Crossed Equator and entered Northern Hemisphere Sunny becoming cloudy Low SE'ly swell
3-10-77	526	05°23N 13°49W	SxW 4	27.5°	26.7°	Vaal Derby Cloudy but clearing Low SE'ly swell
4-10-77	494	12°27N 17°40W	Var. 1/2	27.9°	28.3°	1834 Passed Cape Vert—Most Westerly point of Africa Sunny with light haze No swell The Generation Game S.A. Vaal Flappers
5-10-77	494	20°41N 17°39W	NExE 2/3	21.8°	23.9°	2045 Crossed The Tropic of Cancer and left Tropics Fancy Dress Parade S.A. Vaal Minstrels Sunny with light haze Low N'ly swell
6-10-77	460	27°58N 15°16W	NNW 5/6	22.3°	22.2°	1248 Arrived Las Palmas 1800 Departed Las Palmas 2215 Music From The Shows Sunny with scattered cloud Low N'ly swell
7-10-77	314	32°54N 13°33W	NNW 6	19.5°	22.7°	2115 Olde Tyme Music Hall Sunny with scattered cloud Heavy NNW'ly swell
8-10-77	448	40°02N 10°47W	WSW 5	19.6°	18.9°	2200 Passed Cape Finisterre and entered Bay of Biscay Overcast with scattered cloud Heavy NNW'ly swell Mailship Farewell. Pyjama Party at the Vaal a go-go
9-10-77	482	47°23N 06°27W	NWxW 5	14.1°	14.4°	Divine Service conducted by the Commodore 1621 Rounded Ushant, left Bay of Biscay and entered English Channel. O'cast occl. show's Mod. W'ly swell 2130 Farewell Sing along
10-10-77	294 At Southampton					0330 Expected arrival off Pilot station 0630 Expected arrival alongside berth

TOTAL DISTANCE 5982 MILES - - AVERAGE SPEED 20·33 KNOTS

Other information in Mr. Inkson's Instruction Books concerned working conditions on board "SAXON" in 1925. Standing Orders plainly showed that the Night Watchman had to call 'All Hands' at 4.00am. The daily hours of work, including Sundays, were eleven hours starting from the commencement of their duties and were inclusive of meal times for which they were allowed one and three quarter hours. Certain Leading Hands were allowed to mess in the Third Class Saloon at certain times - Coffee at 6.30am, Breakfast at 8.30am, Dinner at 1.00pm and Tea at 5.00pm. Overtime for Assistant Ratings was then 1s 3d per hour (now 6p).

Strict regulations were laid down regarding the uniform all hands should wear, both dress and undress uniforms being listed. The issue of crew bedding was standardised too. Boatswains and Carpenters were provided with beds, bedding, crockery and glassware. Quartermasters, Painters, Lamp Trimmers and Boatswains Mates had beds and bolsters only. Stewards had bed and bedding. No mention was made of Engine Room staff.

Head Waiters, First Fore-Cabin Stewards and First Steerage Stewards were earnestly enjoined to switch off lights after they were no longer required, to avoid waste of electric power. Lights had to be turned off in the Saloons at 10.30pm in any event and in the Smoke Rooms at 11.30pm.

Bandsmen were allowed to do the work of Stewards when it did not interfere with their duties but no work, such as scrubbing etc, was to be given to them which could injure their fingers. Hairdressers had to be at their appointed places and times in each Class and were only allowed to sell those articles to be of use to passengers during the voyage and approved by the Company. At that time hair cutting was one shilling (now 5p) and beard trimming 6d (3p). Stewardesses had to attend children's meals and another of their duties, particularly in the Third Class, was to see that Ladies were off the decks at a reasonable hour (no mention was made as to what that hour might be).

On the victualling side it was stressed that a liberal table should be kept in all Classes and 'liberality, strict economy and no waste' had to be the watchword. Beef tea was available for First Class Passengers only between breakfast and luncheon, but to others only on application to the Medical Officer or Chief Steward. Chicken broth was only dispensed on the orders of the Medical Officer.

The 1920's instructions stated that strict regulations governed the sale of liquor in the ship's bars. This is the first known reference to bars on the Company's ships and does not state how many there were. "ARUNDEL CASTLE", "WINCHESTER CASTLE" and "CARNARVON CASTLE" certainly had four each - one in each Class to serve wine and drinks in the Dining Saloons and one each for the Public Rooms. They were not 'open' bars and passengers had to order their drinks from waiters and stewards. The "ATHLONE CASTLE", "STIRLING CASTLE" and "CAPETOWN CASTLE" also had a Smoke Room Bar but, again, it was steward service only. It was only in the later years of their service that the First Class Lounge Bars of these ships were extended to form a Cocktail Bar where passengers could sit and order their own drinks. A result of this, of course, was that the barmen not only had to know their job, but had to have a presentable manner and appearance.

The "PRETORIA CASTLE" and "EDINBURGH CASTLE" fared a little better with an 'open' First Class Cocktail Bar and also an 'open' Tourist Class Lounge Bar. It was not until the arrival of "PENDENNIS CASTLE" that 'open' bars were generally available, including a Lido Bar near the swimming pools which proved very popular. The cruise liner "REINA DEL MAR" had a total of ten bars on board including the crew bar.

With passengers now meeting barmen there was often a rapport between them and the passengers would often write to them to tell them when next they would be travelling, often before the Purser's staff themselves were informed. There was one famous titled Lady whose favourite drink was a White Lady cocktail and as soon as she embarked she went straight to the First Class Cocktail Bar to check that the barman knew her requirements, if not she went to the bar at 5.00pm after sailing to give him exact details as to how she wanted her drink mixed. Another story is told that a regular passenger disembarked at Cape Town and travelled to his home in Rhodesia where later he met a friend who said he was shortly travelling to England in the "TRANSVAAL CASTLE". He was told to contact 'Jack the Barman' who would look after him. No surname was mentioned.

After the First World War, Union-Castle set up their own wine and spirit warehouse in the Old Docks Southampton known as 'O' Vault. Whisky, gin and rum were received in huge casks 100% proof and the spirits were reduced to the necessary strengths, bottled and labelled with Union-Castle own labels. They also received South African wine in bulk and this too was bottled and labelled before being shipped on the Company's passenger and cargo ships as required.

In pre- and immediate post-War years the spirits were shipped in two gallon stone jars, encased in wickerwork and sealed. On board members of the Wine Department poured them into four gallon stone jars in the bars which were then locked into a cupboard with just patent taps protruding. These taps were fitted with a register which recorded how many tots had been sold. The keys and opening figures were handed to the Purser who supervised further fillings at sea. At the end of the voyage any remains were measured and returned to the ship's bond room. It was supposed to have been a fool-proof method of supervising the sale of spirits on board, but there were often doubts as to its efficiency. Also at the end of each voyage, all jars and empty bottles were landed to the Empties Store in the Old Docks Office where they were thoroughly washed before being returned to 'O' Vault for future use.

It was said at the time that the bottle washer used to up-end the spirit jars into a container for a few days to thoroughly drain and managed to get a few tots of spirits each week, which no doubt he put to good advantage.

The Vault was eventually demolished to make way for a new car ferry service to the Continent and bottling operations were transferred to the New Docks to be operated by the Ship's Stores supplier, C.G. Hibbert & Company Limited. At the same time the Old Docks Empties Store was abandoned and the bulk spirits came to an end.

The "WINDSOR CASTLE" and "TRANSVAAL CASTLE" were later adapted to ship draught lager in bulk for the ship's use. "WINDSOR CASTLE" had nine huge stainless steel tanks fitted in the storerooms (approximately 360 gallons each) and supply lines, some over 500 feet long, connected the tanks with the bars. The beer was dispensed to the bars by pumps operated by air from a compressor, a meter reading the output, each stroke of the pump measuring one Imperial pint. There was a flash cooler in each bar to finally cool the beer. The lager was delivered alongside by road tankers by the two suppliers - Harp Lager and Allied Brewers - and they pumped over 3,000 gallons of lager into the tanks. To enable the tank gauges to be checked accurately, if there happened to be a list to the ship, an inclinometer was fitted in the tank room similar to the one on the Bridge.

The first Wine Lists, although displaying a full range of wines, spirits, beers, minerals, tobacco etc, were very plain, but in time became more colourful and artistic. Separate Cocktail Lists were also produced.

As for prices, in 1958 whisky was 1s per tot (now 5p), but this was one-fifth of a gill, twice the size of a shore side measure. Gin was 9d (now 4p), best brandy 3s 6d (18p), beer 1s a bottle (5p), fruit juices 10d (4p) and a packet of cigarettes was 4s (20p).

After the sinking of the "TITANIC", when many families lost several relatives, shipping companies decided that, in future, fathers, brothers and sons should not serve in the same ship. This did not, however, stop families from serving for many years in the same Company. Two families in Union-Castle service are good examples. The Gilmore family, whose father was a Head Waiter and mother a Laundress, had three sons who all served in the Catering Department with a family total of seventy-five years.

The best record, as far as I know, is that of the Pyne family. The father, James Pyne, worked in the General Stores in Southampton after leaving the Royal Navy and seven of his sons joined the Company (one retiring as Chief Steward) to be followed by five grandsons and one great grandson. This most remarkable family history adds up to 342 years with the Union Line and the Union-Castle Line.

CHAPTER FOURTEEN

MISHAPS AT SEA

Fortunately, the Union-Castle Line did not lose any ships after the Second World War, but in the early days there were several cases of ships foundering or suffering near disaster. One of the earliest of these was the Union Line's "AMERICAN" in 1880, homeward bound from the Cape just North of the Equator when a propeller shaft broke and the remains bent so that before the engine could be stopped it tore away the ships plating and bulkheads. Water flooded in, but before she sank all the passengers and crew were safely transferred to life boats, but it was several weeks before all the boats were accounted for. One of the boats was rescued by a West African ship, but before she could reach Las Palmas she herself went aground so those unfortunate passengers were shipwrecked twice in one voyage.

The 1,055 ton "CAMBRIAN", also of the Union Line, fared slightly better. When she was built in 1860 her performance was very good, but by 1871 she had become rather slow. She left Southampton in the September of that year very light in draught with no ballast. Due to bad weather she was nearing Cape Town when her coal supply ran out. All available wood was used and even the Bridge was chopped down to use as fuel. The ship managed to creep into Saldanha Bay when the 3rd Officer took horse and galloped into Cape Town resulting in the coaster "NATAL" bringing out a coal supply. The "CAMBRIAN" finally sailed into Table Bay, but was disposed of shortly afterwards, finally being sunk in the bay of Biscay in 1882. Saldanha Bay was a waterless and bleak area and later the same year the "NORSEMAN" spent three weeks there with smallpox on board. The "NORHAM" and "CELT" also went into quarantine there for the same reason.

In 1874 the first "WINDSOR CASTLE" was almost destroyed by fire, but her Captain managed to get the ship into La Coruna in Spain. However, two years later, on 18th October 1876, she hit a reef near Cape Town. Two officers set out by boat to the mainland, hired a horse and cart and rode to Cape Town and the coaster "FLORENCE" was despatched to pick up the passengers and crew. They, in the meantime, had landed to nearby Dassen Island where there were just three Portuguese inhabitants who shipped penguin eggs and guano to the mainland so the shipwrecked passengers and crew ate boiled penguin eggs before pitching camp for the night - not knowing when they would be rescued. The "FLORENCE" arrived the next day, but could not accommodate everyone so some had to stay on the island for another day. Within a week, "WINDSOR CASTLE" broke up on the reef.

Another unusual voyage was that of the Union Line's "SABINE" of 2,992 tons built in 1894 and purchased by the Company in 1898, later being transferred to the Bullard King Line. In 1909 she was chartered to search for a passenger liner named "WATARAH" which had left Durban in the July of that year bound for Cape Town and the UK, but simply disappeared. The "SABINE" was fitted with a search light and towing gear, was bunkered for a long voyage and carried a Royal Navy Officer and several ratings. She had no wireless and nothing was heard of her for three months. Then in the December she returned to Table Bay without finding any trace of the missing ship despite a search covering 14,000 miles. To this day, no further light has been thrown on the mystery.

In December 1919 the mail liner "CARISBROOK CASTLE" (Captain Hoskins) was approaching the entrance to Durban harbour at 10.00am on a fine, clear, calm day when the ship hit a submerged object, heeling violently to port followed by a list to starboard, to the sound of breaking crockery and glasses. She uprighted herself and immediately continued steaming on a steady course into Durban. There it was found that she had about a dozen buckled plates.

It was thought at the time that she had hit the wreck of the "WARATAH", but at a later Court of Enquiry it was decided that the "CARISBROOK CASTLE" was off course and had hit a shoal. The liner was patched up before returning to Cape Town for dry-docking. The vessel was so big her fine counter-stern overhung the caisson of the dry dock and all vehicular traffic had to be diverted.

It was the end of January before she was ready to leave on her homeward voyage, but in the meantime another job was found for her crew. Following a wave of unrest, the native dock labourers were all sent back up country under armed guard and the port came to a halt. However, on arrival of the outward-bound "SAXON" the crew of the "CARISBROOK CASTLE" came to the rescue and in a short time all the cargo and baggage was unloaded. In the three weeks of the unrest, all mail ship and even those on the Mauritius run ("COMRIE CASTLE" and "CLUNY CASTLE") sailed on time thanks to the work of the 'Carisbrook Stevedoring Company' who were amply recompensed.

In 1893 the "HAWARDEN CASTLE" went to the rescue of the "IONIC" which was disabled and took her in tow to Cape Town. Later the crew received £7,000 salvage money.

The same year saw the end of the 2,966 ton "CONWAY CASTLE". For several years she ran as a mail steamer before being transferred to the Intermediate service. In 1892 she was re-engined and refurbished (including the installation of electric lights) and placed on the Mauritius run. However, on 10th May 1893 she ran ashore fifty miles south of Tamatare whilst en route to Durban. The passengers spent ten days ashore under the most primitive conditions at Vatomandry before being rescued by Union Line's "ARAB". "CONWAY CASTLE" herself remained fast on the rocks and broke up within a short time.

In 1922 the homeward bound "KINFAUNS CASTLE" picked up distress signals from the German liner "HAMMONIA" in difficulties off Cape Finisterre, whilst en route to South America with a large number of passengers. Captain E.W. Day immediately changed course and when the "KINFAUNS CASTLE" arrived at the scene lowered life boats. A gale force wind was blowing with high seas making rescue difficult and hazardous and it took eight hours to rescue 340 passengers and crew from the German ship which then sank.

Just after the end of the First World War the "BERWICK CASTLE" was responsible for sinking one of Britain's first submarines, the "Al", near the Nab Tower off the Isle of Wight.

One of the most famous murders took place on "DURBAN CASTLE" in 1947 whilst en route to Southampton from the Cape. At the time she was temporarily on the mail run whilst other ships were being reconditioned after their War service. A young actress was visited in her cabin one night by a ship's steward, it would appear by mutual consent. Whether she died from natural causes or he inadvertently killed her - in a moment of passion - was not clear, but he panicked and pushed her body through the porthole. The incident was even more newsworthy since it was the first time, in British legal history, of a murder trail without a body being produced. The steward was found guilty and sent to prison.

Just after the second War a Catering Department storekeeper, who it was believed had many domestic problems, jumped overboard from a mail ship just prior to docking at Southampton. Sad as it was it caused many problems for the Southampton staff since he took all his books and records of the voyage with him.

As with all ships there have been many incidents during the years of people falling overboard, some accidentally and some deliberate - watching the sea rushing by can have a strange affect on some people. In 1976 a sixty-four year old woman fell overboard from the "WINDSOR CASTLE". When she was reported missing an hour later the ship was searched and eventually turned round when Captain P.St.Q. Beadon retraced his course and within twenty minutes the lady was spotted close by and swimming strongly. A life boat was lowered and she was rescued.

In 1973 a deck hand on "S.A. VAAL" saw a lady passenger go overboard, he alerted the Bridge, a life boat was launched and the unconscious lady recovered and the liner continued on her way. The whole incident took just fifty-one minutes.

A more amazing rescue took place in 1970 by the same ship when a young man went overboard and, at first, became unconscious after hitting the water. However, he was strong and a good swimmer. When he was reported missing and the ship had been searched, Captain A.A. Freer turned the ship and plotted a forty mile search area. Eventually the man was spotted only eighty yards off course, still swimming. He had been in the water for eleven and a half hours and was suffering from shock and exposure, but made a marvellous recovery. On arrival at Southampton the press described him as the luckiest man alive.

In November 1967 the Shaw Savill liner "NORTHERN STAR" was berthed at Durban when a electric fire broke out and her passengers had to be evacuated. "WINDSOR CASTLE" was berthed nearby discharging cargo and the passengers were invited on board and given refreshment and cabins for the night. At a ceremony on board "WINDSOR CASTLE" at Southampton the following August, Shaw Savill presented a painting to Mr. B.G.S. Cayzer, Deputy Chairman, to mark their appreciation of the Company's help.

Commodore R.M. Wright was on "WINDSOR CASTLE" in March 1972 when he received a request for urgent medical assistance from a Danish tanker after an apprentice had been injured. Captain Wright headed for the tanker and, on arrival, Dr. McNamara and his hospital attendant boarded the tanker by the Pilot's ladder. He decided to return to the liner with the patient for further examination so they all had to reboard the life boat. This was no easy task since the life boat was rising and falling ten feet alongside the massive side of the tanker. However, the injuries proved not so bad as at first thought and the apprentice was flown home to Denmark from Las Palmas.

In 1960 a tragic accident happened on board "CAPETOWN CASTLE", northbound from the Cape, as she was approaching Las Palmas. An explosion occurred in the Engine Room and seven engineering officers and ratings were killed. Tugs came out from Las Palmas to help put out the fire and to tow the ship into port. Passengers were brought home by other ships and "CAPETOWN CASTLE" herself eventually arrived back in Southampton, but was out of action for some time and an Intermediate vessel was brought round from London to take her place. Among those killed was Mr. Stan Logan, Chief Engineer, who, although injured himself, made many valiant attempts to save his colleagues.

From time to time fire exercises were held in Southampton Docks involving Union-Castle staff and the local fire brigade to practice fire fighting on board ship and to evaluate the stability of a ship if large quantities of water was pumped in during the course of a fire.

One such exercise was held in "S.A. VAAL" early in 1968. Shortly afterwards this was put into effect for real. On 27th May of that year "PENDENNIS CASTLE" arrived in Southampton from the Cape and as soon as the passengers had disembarked workmen commenced a conversion and refurbishment of First Class cabins and suites. During the evening someone caused a spark and fire spread to the whole area. The Southampton Fire Brigade arrived within minutes and the fire was put out. Beside fire damage there was also more damage caused by the dense smoke. Despite this the liner still sailed on her next voyage on time, although a few cabins were out of use.

Another ship to go to a rescue was the "BLOEMFONTEIN CASTLE" on 8th January 1953 with 500 passengers on board. She was en route from Lourenco Marques to Beira when an S.O.S. was received from the Holland Afrika liner "KLIPFONTEIN" that she was sinking and passengers and crew had abandoned ship. The two ships had been loading at adjacent berths at Lourenco Marques, but the Dutch ship had sailed earlier and was now several hours ahead.

Captain J.A. Ferguson on "BLOEMFONTEIN CASTLE" immediately ordered maximum speed and within two hours sighted the other ship well down by the bows and with all the life boats pulling towards their rescuers. The liner lowered her motor boat which towed in the life boats and, despite a heavy swell, embarked 234 passengers and crew from the stricken ship. The Smoke Room was turned into an emergency first-aid post complete with hot refreshments and blankets. In the meantime the "KLIPFONTEIN", which had struck an emerged object, shot stern first in the air and sank. Accommodation for the night was a problem on an already full ship, but was resolved until docking at Beira the next day when the survivors were landed.

CHAPTER FIFTEEN

PEOPLE AND SHIPS

Towards the end of the Company's history many older members of the staff, ashore and afloat, tended to write to the Group's House Magazine, the "Clansman" with interesting anecdotes of by-gone days and some of these are worth recalling.

It seems that Union-Castle liners have always figured prominently among Merchant Navy ships in attendance at the Sovereign's Reviews of the Fleet at Spithead. In 1897 "ARUNDEL CASTLE" was present at Queen Victoria's Diamond Jubilee Review, and at King Edward VII's Review in 1905 the "ARMADALE CASTLE" played host to the French Fleet. In 1935 the "WARWICK CASTLE" took a large party of Company's guests to King George V's Silver Jubilee Review and the Master, Captain W. Morton-Betts was invited on board the Royal Yacht and was presented to His Majesty. At Queen Elizabeth's Coronation Review in 1953, the "PRETORIA CASTLE" left Southampton for Spithead and among those on board were Queen Salotti of Tonga and her party.

The Union Line's "NORMAN" was the first ship to give a cinematograph show on board. Carl Henty, an early enthusiast, brought his equipment on board and gave a demonstration. Earlier, in November 1894, the same ship took out to South Africa the touring company which was due to open the 'Empire Palace of Varieties' in Johannesburg.

At the beginning of the Century came the exciting advent of wireless and Mr. John Brinton, a Director of Donald Currie & Company, attended an early demonstration at the offices of Marconi & Company. He was impressed and asked if one of their ships could be reported passing one of the Marconi stations. Accordingly the passage of the "CARISBROOK CASTLE", outward bound for the Cape, was reported by the Alum Bay station on the Isle of Wight and a wireless message was sent to Bournemouth where it was passed on to Brinton by land-line. The first of the Company's vessels to be fitted with wireless was the "BALMORAL CASTLE" for her maiden voyage in 1910 and by the end of the following year all Mail and Intermediate vessels had been so fitted. After completing two voyages to South Africa the "BALMORAL CASTLE" made a few cruises round the Isle of Wight carrying guests specially invited by the Company and on one of these cruises was Marconi himself who delighted his fellow guests with details of his invention and many used it for the first time. The cost was then 10½d per word.

An incident which was widely reported by the media at the time was the "CAPETOWN CASTLE" gold robbery. In February 1965 the vessel had just arrived at Southampton and, as usual, the first cargo to be discharged (even as the passengers disembarked) was the gold bullion which was stowed in a special specie room forward. It was then normally loaded into special wagons and taken to Waterloo Station in London, under guard, where it was collected by the Bank of England. On this occasion, however, it was found that twenty gold bars were missing despite having been correctly loaded in the Cape.

An announcement was made on the ship's tannoy that nobody was to be allowed ashore and that the ship was to be thoroughly searched. Shortly afterwards many splashings were heard in the sea alongside and the Chief Officer lowered a boat to investigate, but it was discovered that some members of the crew were just dumping illegal possessions.

During the days which followed, squads of police and customs officers continued their search of the ship and one squad in particular searched the beer room where every case of beer and minerals was moved (including hundreds of cases recently shipped in south Africa) and also the 'dry' storerooms where every sack of potatoes and flour and all the tinned goods were moved, all to no avail. It was hard and heavy work, but a bottle or two of beer apiece soon relieved the pain.

However, despite the searches and other enquiries by Chief Inspector Bertram Adams (shortly due to retire) and Sergeant McCullough of the Southampton Police, nothing more was discovered and the ship sailed on her next voyage on time. However, the police persevered and more information came to light with the result that two voyages later, the two police officers flew to Durban, boarded the ship and the missing gold bars were found embedded in concrete in a sand box at the foot of the forward mast. Some members of the crew were arrested and flown back to England where they were convicted and sent to prison.

In April 1875 the President of the Transvaal sailed to England in the "WALMER CASTLE". Sir Donald Currie had obtained the flag of the Transvaal and four days after sailing from Cape Town, when the President celebrated his birthday, the flag was flown from the mast head to the delight of the President.

In February 1963 the "PRETORIA CASTLE" had twenty-four hours of glory when Vice-Admiral A. FitzRoy-Talbot CB, DSO, was appointed Commander-in-Chief of the Royal Navy Atlantic and South America stations and sailed in the liner to Cape Town. As her Master, Captain J.P. Smythe RNR, described the event it was tradition that when a shore-based Commander arrives at his station his flag is flown for the first twenty-four hours in the senior Royal Navy ship of the station. As it happened, on the day that Admiral Talbot was due to arrive all R.N. ships were away from South African waters.

After consulting Admiral Talbot and the Admiral Commanding Reserves, the Admiralty gave permission for the "PRETORIA CASTLE" to fly the Admiral's flag on arrival at Cape Town. This was the first occasion in peace-time that a merchant ship was given the privilege of becoming the flag ship of a Naval Commander-in-Chief.

At 8.00am on 21st February, in Table Harbour, the Admiral's Flag was broken out at the main mast, a Blue Ensign at the ensign staff and a Union Jack at the jack staff forward. These were in addition to the South African National Flag and the House Flags normally displayed.

After receiving various local Civic and Naval dignitaries on board and holding a press conference the Admiral, in full ceremonial dress complete with sword, disembarked and was piped over the side by two ex-Navy deck hands.

Later that day a Colour party and guard from H.M.S. Afrikand (Naval shore base) boarded and at sunset the Blue Ensign was hauled down with the due honours usually accorded to the White Ensign only. The Admiral's Flag remained floodlit at the mast head all night until

BRAEMAR CASTLE

struck at 8.00am the next morning when the "PRETORIA CASTLE" resumed her normal duties.

On 1st June 1963 it was the Army's turn to salute the "BRAEMAR CASTLE". On board was Lt. General C.H.P. Harrington CB, CBE, DSO, MC, the new Commander-in-Chief Middle East forces travelling to Aden to take up his new command. Having been contacted the day before by an R.A.F. Shackleton, the "BRAEMAR CASTLE", looking spick and span and dressed overall, was welcomed outside Aden by the aircraft carrier "H.M.S. ARK ROYAL" and destroyers "H.M.S. CORUNNA" and "BLACKPOOL" who escorted the liner to the port limits. "H.M.S. ARK ROYAL", Captain Pollock RN and with Vice-Admiral Gibson flying his Flag, steamed down the port side and made a tight turn before steaming back alongside the starboard side with a fly-past of R.A.F. jets - quite a spectacle for the passengers lining the deck. At the limits of the port an R.A.F. launch and helicopters provided the escort for the last part of the journey.

An interesting booklet on the Union-Castle Line was published in South Africa called "Now and Then" and referred to the breakdown of the crew of an average mail liner as being fifty in the Engine Room, ninety in the Deck Department and 260 in Catering. The members of the crew whom the passengers usually met were the Purser's staff, stewards, barmen, liftmen and medical staff. The article pointed out that they did not often see the chefs, cooks, bakers, storekeepers, painters, plumbers, carpenters or laundry staff. It was only when passengers asked to see the kitchens that they began to realise the activity behind the scenes.

It is worth noting how some ships in the past have had careers lasting many years and sometimes passing through several ownerships. As mentioned earlier "DUNNOTTAR CASTLE", still cruising, is over fifty years old with three owners. The "CELT" was the sixth of the Union Line's original fleet and was built in 1855, but only served the Line for five years before being sold and was finally scrapped after thirty-six years afloat. The "DUNROBIN CASTLE" served for thirty-nine years and the "WARWICK CASTLE" of 1867 had four owners during her forty-nine years at sea.

RAGLAN CASTLE

The "RAGLAN CASTLE" was built in 1897 as an extra Intermediate ship carrying only 135 passengers, but with a considerable cargo capacity. On the outbreak of the Russia-Japanese War in 1905 she was sold to be a store ship for the Russian fleet. Thereafter, she had seven changes of name and even more owners and was capsized in 1929, but was salvaged and continued trading as a whale factory ship.

Greenoch Dockyard Company Limited gave an interesting insight at the behind the scenes activities at a launching ceremony. Despite more modern trends of automatically projecting a bottle of wine onto the bows of the ship, the Dockyard preferred the personal touch whereby the Lady Sponsor smashed the bottle against the ship by hand.

First, the bottle of champagne was dressed with thirty-three feet of ribbon and white satin, most painstakingly arranged and above all, to ensure that there was no chance of fragments of glass showering the official guests. Then the bottle was covered with red, white and blue ribbons, some heather for luck, and ribbon in the colours of the Company. This was secured by tartan ribbon according to the name of the ship (in the case of Clan Line ships) or the ancestry of the Sponsor. The rope from which it was suspended was covered with white satin. A duplicate bottle was also prepared in case of a prior accident to the first. These two bottles were prepared by the two senior typists.

As the guests assembled on the platform, three other typists were hidden under the platform. The launching went ahead and the ship duly named and the bottle of wine smashed against the bows of the ship after which the remains and the rope were dropped to the ground. The three ladies, prepared for a shower of wine, gathered the ribbons etc and quickly returned to the typing pool for a race against time.

Whilst the guests watched the launching before returning to the Company's office for a reception, the girls had fifteen minutes to arrange the heather, ribbons etc into forty to fifty favours which were then handed to the lady guests as they entered the reception. This was always achieved, despite the hazards of broken glass. As a tradition, the young boy who presented the bouquet to the Lady Sponsor at the launching was usually given the cork as a souvenir.

CHAPTER SIXTEEN

PROGRESS IN DESIGN

Although only steamships have been employed on the mail run, the earlier ships were also equipped with sails in case of engine failure and to conserve fuel. The Union Line gave a bonus to Commanders who came home with small coal bills so when the wind conditions were favourable they would 'feather' the propellers, shut down engines and carry on under sail.

Over the years, the design of shipping has progressed from multi-funnels and masts to the more modern approach of an almost invisible funnel and no mast. The 613 ton "PHOEBE" in 1858 had a single 'woodbine' funnel, but in addition had four masts carrying eight sails. Even in 1880 the 3,507 ton "KINFAUNS CASTLE" (the first ocean-going steamship constructed of steel not iron) sported two masts and a single funnel. It was another ten years before sail was discontinued, but masts remained (mainly for navigational and signalling purposes) and tended to grow in number.

The second "ARUNDEL CASTLE" in 1895 had but one funnel, but four masts and the "DUNNOTTAR CASTLE" in 1890 two masts and three funnels. However, as the number of masts decreased the number of funnels increased, culminating in the Union-Castle Line with the "ARUNDEL CASTLE" and "WINDSOR CASTLE" in 1921 with four funnels and two masts and a cruiser stern (two of the funnels were later removed). In more recent times, "SOUTHAMPTON CASTLE" and "GOOD HOPE CASTLE" had their mast based in the funnel.

Incidently, a novel design was incorporated in the American cruise ships "ARGENTINA" and "BRAZIL" with a single wide funnel which also incorporated the Captain's cabin and a Sun Lounge.

Conditions for passengers in the early days were almost primitive. There were just salt-water tubs provided on deck for the gentlemen between 5.00am and 7.00am when the decks were being washed down and it was not until the "CAMBRIAN", in 1860, that a bathroom was provided. At first there was little lingering in the Saloon after a meal, since the Dining Saloons were in the centre of the ships with hard benches and so cramped, passengers were glad to leave, especially in bad weather. Some hardy passengers however, gathered there in the evenings to play cards since there was no Lounge or Smoke Room. Smoking was only allowed in the day-time on certain parts of the open deck, in case of fire below decks which was always an ever present danger. Illumination for passenger cabins were candles conveniently placed between cabins, so that one candle lit two cabins. This was also easier for them to be extinguished at 10.30pm.

The 1860's saw a slight improvement with the provision of a small Ladies cabin in their stern and with each ship carrying a doctor. Breakfast was at 9.00am, Luncheon at Noon, Afternoon Tea was served at 3.00pm and at 4.30pm sounded a dressing bell for Dinner at 5.00pm. All lights had to be extinguished by 10.30pm.

Conditions gradually improved and the "GERMAN", in 1877, had a large dining room complete with piano so that passengers could entertain themselves. The Saloon was also used for reading, writing and whist and when the weather was fine there were deck games available and the Quarter Deck prepared for dancing.

The Chairman of the Line was once heard to say that even the Second Class passengers were wanting baths, a piano and ice in their drinks. A few years later refrigerators were provided in the First Class and the Second Class passengers got their ice. The next stage was the installation of electric lights throughout the ships, more comfortable accommodation and the provision of a Smoke Room for the men. Ladies were not admitted and the lights could remain on until 11.00pm. By the advent of the "BALMORAL CASTLE" in 1910, passenger accommodation was vastly improved with a spacious First Class Dining Saloon, a Lounge, Music Room and a Library.

The arrival of "ARUNDEL CASTLE" in 1921 represented the next major change in the design of the Company's mail ships. For a start she was half as large again as the "BALMORAL CASTLE" and among her new facilities were single-berth cabins, cabins with bathrooms and a new system of forced draught throughout the ship in place of individual fans. There was also a built-in swimming pool, a gymnasium and a lift for passengers. The Promenade Deck was also enclosed with glass windows instead of canvas wind-breakers. With a certain amount of relaxation following the First World War, ladies often travelled alone and they were also permitted to use the Smoke Room.

The next stage in development was the launch of the "CARNARVON CASTLE" followed by "WINCHESTER CASTLE" and "WARWICK CASTLE". Apart from being the first ships in the fleet over 20,000 tons, they were also the first of many ships powered by diesel engines. They all had elegant First Class Lounges with oak panelled Smoke Rooms. In the First Class Dining Saloon was a minstrel's gallery for the ship's orchestra with a swimming pool underneath.

Over the years the words Third Class were gradually changed to Tourist Class and the Second Class merged into the First Class to provide a greater range of fares. The "STIRLING CASTLE" in 1936 used Cabin Class instead of Tourist with the "BLOEMFONTEIN CASTLE" and the later Intermediate ships being all one-class.

The next and final change in ship design came with the launching of "PENDENNIS CASTLE", "WINDSOR CASTLE" and "TRANSVAAL CASTLE" when Union-Castle ships really did enter the realms of luxury travel with everything on board a passenger could want.

By the 1890's South Africa was developing into quite a tourist attraction, but the hotel accommodation was very inadequate, especially for the well-to-do classes who travelled First Class on mail ships. Sir Donald Currie was quite concerned over this and bought a large estate to the north of Cape Town called Mount Nelson. He formed a new company called African Lands and Hotels Limited and set about building a hotel on the site. He engaged the best local architect and a well known firm of London builders to build and create a most luxurious hotel which he named Mount Nelson Hotel. His son-in-law, Mr. F.J. Mirrieless, went to Switzerland to engage a top class Chef and other staff and on 6th March 1899 the hotel received its first guests. Since then it has been enlarged several times and now stands in its own beautiful gardens and has become known throughout the world for its luxury and splendour.

When the 530 ton "DANE" made her first voyage to the Cape she only carried cargo to the value of £102 due to her hurried departure. Since then cargo has been referred to by tonnage rather than value. As word of the new service to South Africa spread so more cargo was forthcoming and over the years it has comprised of a whole range of goods, from locomotives and machinery to clothes etc. This has been in addition, of course, to the mail. Some of the extremely heavy lifts, in the later years, such as locomotives and buses were shipped in the general cargo ships such as "DRAKENSBERG CASTLE" and "KENILWORTH CASTLE". Another feature was the carrying of live stock to the Cape and valuable race horses.

The biggest problem has always been to find sufficient cargo for the homeward journey - for many years 78% of the cargo carried in the Company's ships comprised mainly gold and diamonds.

At the request of the South African fruit farmers the "GRANTULLY CASTLE" in 1888 was fitted with a small refrigerated chamber of 30 tons capacity, the insulation being of cork. The following February she sailed from Cape Town with 320 boxes of grapes, the first experiment in handling fruit, but it was in such a poor condition on arrival that the experiment was abandoned.

Sir Donald's son-in-law, Mr. P. Molteno, took a considerable interest in the fruit trade and persuaded both Company's to fit some vessels with cold chambers. The first consignment of peaches arrived in the UK on "DRUMMOND CASTLE" in February 1892 and a few days later were being sold in London. Cold storage capacity was increased and by the following year 102 tons of fruit was exported to Britain. However, further problems had to be overcome including transportation from the farms and cold storage in the docks at Cape Town.

This resulted in fruit farming being given a boost and the value of land rose. Even Cecil Rhodes bought fruit farms at Franschoek and Drakenstein. In time, of course, more refrigerated cargo space was made available in the Mail and Intermediate ships, which was followed

"WINDSOR CASTLE" sailing from Southampton passing "GOOD HOPE CASTLE" at the inward berth and five 'R' ships laid up pending the start of the fruit season.

by the introduction of the 'R' Class ships with a carrying capacity of 8,000 tons of fruit.

Some of the general cargo ships often carried more exotic cargoes including various unusual fish for aquariums in Britain. In 1947 the "KENILWORTH CASTLE" brought over for London Zoo a valuable collection of wild animals including hippos, lions, buffaloes, rhinos and an 150 year old giant tortoise.

For many years South African wine had been exported to the UK in barrels and on arrival they were sent to the warehouse of the South African Wine Farmers Association. They later moved their depot to new premises in Southampton New Docks for their warehousing, bottling and distribution operations. Following this massive 60,000 gallon tanks were fitted in the holds of "WINDSOR CASTLE", "SOUTHAMPTON CASTLE" and "GOOD HOPE CASTLE" and the wine shipped in bulk and pumped ashore at Southampton to road tankers for the short journey to S.A.W.F.A.'s premises.

CHAPTER SEVENTEEN

THE CLAN LINE INFLUENCE

The formation of the Clan Line was the result of the aspirations of two men. Alexander Stephen was the senior partner of Alexander Stephen & Sons who owned a well equipped ship building yard at Govan on the Clyde. Ship building was in the doldrums in 1877 and he was anxious to get new orders when he was introduced to Charles Cayzer who had earlier joined the British India Steam Navigation Company in Bombay, and now had the idea of starting his own shipping line from Glasgow and Liverpool direct to Bombay. Cayzer had difficulty in getting financial support and eventually Stephen decided to back him himself, with some help from his friends. Thus the Clan Line was formed.

Whilst retaining his office in Liverpool, Cayzer decided to move his head office to Glasgow. He chose as his partner to look after the Liverpool office a British India ship master named Alexander Irvine who unfortunately died the following year, but the holding company has always been known as Cayzer-Irvine & Company.

The venture got off the ground and his first steam ship was called "CLAN ALPINE" to be followed shortly afterwards by the "CLAN FRASER", "CLAN RANALD", "CLAN GORDON", "CLAN STUART" and "CLAN LAMONT". There then entered Sir John Muir Bt, a senior partner in James Finlay & Company with Agencies in Bombay and Calcutta and he succeeded in getting financial backing for a new Fleet, the Clan Line Associate Steamers, to be managed by Cayzer-Irvine & Company. In 1881 they ventured into the South African trade with the "CLAN LAMONT" arriving in Cape Town. In 1887 Charles Cayzer was created a Knight and in 1904 a Baronet, taking the title Sir Charles Cayzer of Gartmore.

All was now going well for him. He was operating twenty-five ships on three profit-making routes: Lancashire cotton to India, importing tea and coffee from the Malabar Coast, dates from the Persian Gulf, carpets from Madras and sugar from Mauritius. In conjunction with the Union Line a service from North America to South Africa was established in rivalry with the Castle Line. Clan Line activities took their ships to Indonesia, Burma, Australia and New Zealand, from the last two countries were imported meat and dairy produce so like the Union-Castle Line they invested in refrigerated cargo ships.

Cayzer's two sons were learning the trade and on the death of Sir Charles Cayzer in 1916 his eldest son, now Sir August Cayzer, became Chairman of the Company. In the First World War Clan Line suffered the loss of twenty-eight ships and 300 men. These ships were very quickly replaced.

After the War, despite his Father's previous loyalty to Glasgow, he moved the head office to London although he kept on the office in Hope Street, Glasgow. Sir August then made three important purchases, namely the Shire Line, British & South American Navigation Company Limited (opening up the south American trade) and the Greenoch Dockyard Company, where all future Clan Line ships were built.

At this period a severe worldwide depression set in, but the Clan Line were determined not to lay up ships and they were able to balance their trade with India with that of South Africa. Apart from the Union-Castle Line there was further competition on the South African route from the Thomas Steamship Company whose main interest was galvanised iron, causing Clan Line to lower their freight rates to compete. After two years however, the company became insolvent and the Clan Line bought two of their ships at bargain prices.

However, Clan Line prospered and whilst the Lancashire trade faded with new mills being opened in India, other trades flourished and the Company began to be well known for its ability to ship heavy loads from cars to locomotives. Another speciality was the carrying of live stock and they were able to ship consignments of wild animals to the world's zoos. The Greenoch Dockyard Company built two new motor ships which were not very successful and built future ships with twin-screw triple expansion engines capable of a speed of 17½ knots.

In the Second World War all the Clan ships were requisitioned and since many still had coal bunkers as well as fuel tanks, they were capable of sailing longer distances without refuelling. On one voyage a Clan ship steamed 20,000 miles without mishap from Cape Town to Magellan Straits, Panama Canal, North America and Iceland before docking at Methil, a coal port in North Wales. During the War the Clan Line lost thirty-six ships involving 600 men.

Sir August Cayzer retired in 1943 after twenty-five years as Chairman and died a month later. He was succeeded by the Deputy Chairman, Lord Rotherwick, a former Member of Parliament before being elevated to the Peerage. After the War, he bought a number of ships from the Government which proved to be the nucleus of his post-War fleet.

However, the pattern of trade changed again. The emerging nations of the Commonwealth wanted to encourage their own national merchant fleets. The partition of India meant another loss of trade and the sugar importation from Mauritius fell through. In the 1950's the Directors of Clan Line felt that they had to decide upon a new course of action. For some time there had been ideas of taking over the Union-Castle Line and there was no doubt that their fortunes in the South African trade were bound together.

Then in 1954 occurred the death of Sir Vernon Thomson who had been appointed Deputy Chairman in 1932 and Chairman in 1939. He had helped to bring the Company out of the chaos of the Royal Mail takeover and brought the Line back to stability. He was very highly regarded by his staff and had even started pension schemes for them. He was succeeded by Sir George Christopher, Deputy Chairman.

Lord Rotherwick and his Board of Directors felt that the time was now ripe and negotiations were started between the two Companies. However, a group of financiers and ship owners took legal steps to prevent a merger, based on the fact that a clause in the Articles of Association of Union-Castle laid down as a matter of cardinal principal that the Company must be British owned whilst no such clause was part of the Clan Line's Constitution. The judge gave his verdict in favour of the two companies continuing negotiations. Despite certain shareholders still trying to object they eventually agreed terms in favour of the merger.

A new company, British & Commonwealth Shipping Company Limited, was Incorporated on the 1st November 1955 and by the 31st December of that year all Preference and Ordinary Shares in the Union-Castle Line and the Clan Line were acquired by B. & C. and the two Companies merged. Shortly afterwards Sir George Christopher resigned. Chairs of both Companies were held by Lord Rotherwick - on the death of Sir August Cayzer his son, Nicholas, inherited the Baronetcy and now Sir Nicholas (later Lord) Cayzer became Deputy Chairman.

On the death of Lord Rotherwick in 1958, Sir Nicholas took over as Chairman of the Group, the Deputy Chairmen being his brother, Mr. Bernard Cayzer who ultimately supervised passenger and catering services and the decor and lay out of the vessels and Lord Rotherwick's younger son, The Honourable Anthony Cayzer who looked after publicity and cargo matters.

Thus by 1956 the history of Union-Castle Line had entered another phase.

CHAPTER EIGHTEEN

BRITISH & COMMONWEALTH

On the 31st December 1955, at the time of the merger, the new Group owned a total of ninety-nine ships with a further nine being built. The fleet consisted of:

Clan Line		54	
Shire Line		3	= 57
Union-Castle Line	Mail	8	
	Round Africa	7	
	'R' Ships	7	
	General Cargo	5	
	Bullard King	4	
	King Line	11	= 42

This amounted to a total of 942,045 gross registered tons. The ships kept their individual colours, but there were now two House Flags, the B. & C. being flown at the top of the mast over that of the Company concerned.

In his statement at the Annual General Meeting of the British & Commonwealth Shipping Company Limited in July 1959, Sir Nicholas Cayzer referred to the depression in which shipping in general was suffering with nine million tons laid up worldwide. Britain's share was 1.3 million tons, mostly in general cargo and tankers.

Shortly afterwards, five of the Group's ships were scrapped - "GOOD HOPE CASTLE", "DRAKENSBERG CASTLE", "CLAN KENNEDY", "CLAN FORBES" and "KING EDGAR". The Round Africa liner "BLOEMFONTEIN CASTLE" was sold to the Chandris Line who adapted her for cruising and was renamed "PATRIS". Similarly the "DUNNOTTAR CASTLE" was sold to the Incres Line, renamed "VICTORIA" and, for several years afterwards, called in at

ROTHESAY CASTLE

Southampton Water whilst cruising before she also was sold to the Chandris Line.

However, more tonnage was ordered, two new 'R' Class refrigerated cargo ships. The first, "ROTHERWICK CASTLE", was the first Union-Castle ship to be built by the Greenoch Dockyard Company and was launched on 19th August 1959 by Lady Rotherwick, to be followed by the "ROTHESAY CASTLE" on 6th June 1960. They were both of 10,000 tons and had accommodation for twelve passengers (this being the maximum allowed under Board of Trade regulations without further services being provided, such as qualified medical assistance etc).

It was then the Board's aim to reduce the journey time of the mail ships to eleven and a half days which required seven ships to maintain a speed of 22½ knots, but only the "PRETORIA CASTLE" and "EDINBURGH CASTLE" of the present fleet could maintain that speed so a new building programme was laid down.

The Management decided to incorporate numerous additional features to the already ordered "PENDENNIS CASTLE" to give her the additional speed and to fit Denny Brown stabilisers. This necessitated lengthening the ship after the keel had been laid and construction started from 784ft to 764ft, a most unusual operation which was successfully completed. The vessel was launched by the Dowager Lady Rotherwick in Belfast on the 24th December 1957 and an excellent portrait of her was displayed on the main staircase leading down to the First Class Saloon.

The "PENDENNIS CASTLE" was of 28,582 tons with accommodation for 197 First Class passengers, 539 in the Tourist Class and a crew of 419. This was the first time that a ship of this name served in the Merchant Navy, but behind the Captain's table in the First Class Dining Saloon hung a portrait of Admiral Lord George Churchill who commanded an 18th Century warship of that name.

Apart from the stabilisers there were other innovations. She was the first Union-Castle liner to have a Private Dining Room, adjacent to the First Class Dining Saloon, where passengers could hold private parties to entertain their guests. Both swimming pools were on the upper decks with Verandah Cafes and Lido Bars nearby. Air-conditioning was widely used and telephones installed in all First Class cabins for cabin service, inter-cabin use and also for ship-to-shore calls via the Radio Room.

The general furnishing and decoration of the passenger accommodation had been supervised by Mr. Bernard Cayzer with the help of the design consultant, Miss Jean Munro. One of the works of art featured was a large screen in the First Class Lounge made of filigree glass. The (by then) traditional Long Gallery had disappeared and all the Public Rooms rearranged. Medical facilities on board were also improved.

The "PENDENNIS CASTLE" was also the first Union-Castle ship to have 'suites' whereby passengers could have a day-room, bedroom, bathroom etc in a combined area. There were also cabins-de-luxe where a day-room was converted into a bedroom with separate bathroom facilities.

She was warmly welcomed at Southampton and various parties of travel agents and shippers were shown her many facilities before she sailed on her maiden voyage on 1st January 1959 under the Command of the new Commodore, Captain G.H. Mayhew, who only that morning had been awarded the CBE in the New Year's Honours List. Many of the officers who sailed on that voyage were later appointed to senior positions in the Company.

A pamphlet issued to passengers described the various flags flown during the voyage. It explained that the Red Ensign (the flag of the Merchant Navy) was always flown from the ensign staff at the stern at 8.00am to sunset. The House flags were flown from the starboard yard of the signal mast when entering or leaving harbour. At the top of the signal mast was flown the National Flag of the country when entering port and at the same time a small Union-Castle flag was flown from the jack staff in the bows. A red and white flag indicated that a Pilot was on board.

A Royal Mail flag (white with the words 'Royal Mail' in red) was displayed at the port yard of the signal mast. A yellow 'Q' flag meant 'I have a clean bill of health and request pratique' and was flown until the Port Doctor had boarded and cleared the ship. It was also an indication to the Customs Authorities that the ship had arrived

WINDSOR CASTLE

from foreign waters. A 'P' flag (white with blue edging and also referred to as the Blue Peter) was flown on the day of sailing (or twenty-four hours in advance) to give warning that the ship was about to sail.

Unfortunately, as is often the case, welcoming a new liner also meant saying farewell to another - this time it was the thirty-eight year old "ARUNDEL CASTLE" which had completed valuable service to the Company. On her last departure from the Cape - her 211th voyage - she was dressed overall and given a rapturous farewell watched by enormous crowds on the quayside. She eventually sailed from Southampton on the 30th December 1959 for the Far East to be broken up and her sailing was watched by members of the staff, former crew members and local towns folk with much sadness.

The first mail ship to be actually ordered by the British & Commonwealth was built by Cammel Laird at Birkenhead. She was launched on 23rd June 1959 by Her Majesty the Queen Mother and named "WINDSOR CASTLE", the third ship to bear that name. At 38,000 tons she was then the largest ship to be built on Merseyside, the largest to be built in Britain since the "QUEEN ELIZABETH" and certainly the largest to be built for Union-Castle Line. It was also the first launching of a British Merchant ship to be broadcast 'live' on television. Her accommodation was designed for 239 First Class passengers, 591 Tourist Class with a crew of 470 including thirty women. However, one section of cabins were inter-changeable between the two Classes, according to demand.

In his statement at the Annual General Meeting of the Group in July 1960, Sir Nicholas Cayzer said "our new "WINDSOR CASTLE" is complete and we are very proud of her." Among the various official who, in the course of their duties, board all ships in Southampton Docks it was their unanimous opinion that the liner was one of the most luxurious using the port comparing very favourable with the "QUEEN ELIZABETH". Once again, the decoration and design of the accommodation had been supervised by Mr. B.G.S. Cayzer and Miss Jean Munro.

Both First and Tourist Class Lounges faced forward, and one of the most impressive displays was a glass aviary in the First Class

First Class Dining Room - WINDSOR CASTLE

Lounge. A superb Private Dining Saloon had been installed adjacent to the First Class Saloon and the decor, special cutlery, glasses etc would not have disgraced a Royal occasion.

The First Class Dining Saloon itself was one of the most amazing rooms on board. There were only imitation port-holes lit by concealed lighting, the rest of the Saloon in daytime being lit by normal overhead lighting. Opposite the flying staircase into the Saloon and behind the Captain's table was a remarkable painting by Felix Kelly depicting the main drive up to Windsor Castle in the background. In the foreground were two deck chairs which looked as if they had been recently occupied together with some books and glasses. Halfway up the drive were two figures. The tables in the Saloon were on two levels, a well and a slightly raised gallery.

When the passengers came down to Dinner in the evening, the effect had been changed dramatically. Purple curtains covered the imitation port-holes, another purple curtain covered the main mural and disclosed on, either side, two small paintings of the Castle in moonlight. The only other illuminations were table lamps with red shades on each tables. These changes were made electrically and often the Head Waiter ('Dusty' Miller) demonstrated the changes just by pressing a number of buttons on a control panel. The whole process took no longer than a minute and was very impressive.

The decor and accommodation in the Tourist Class was far superior to that in the First Class of some of the more recent ships. The Smoke Room was particularly outstanding. The long room was divided into three sections, the first was the Cockpit Bar with a nautical theme, in the centre was the Rotundra a circular section with a devise called a stovescope, the third section being a comfortable sitting out area. For those able to move from one class to another, each Class was distinguished by different coloured carpeting.

Each Class had a Verandah Cafe and Lido Bar adjacent to the open air swimming pool and every First Class cabin was an outside cabin with a telephone. There were ten de-luxe cabins plus several Suites, each with its own name. Other features included a Health Spa and an Elizabeth Arden salon in addition to the ship's shops.

First Class Lounge - WINDSOR CASTLE

It had been the intention of the Company to have a shake-down cruise from Birkenhead, up the West coast of Scotland and through the Pentland Firth to Leith and then on to Amsterdam before proceeding to Southampton to prepare for her maiden voyage. However, a dock strike intervened and in the end the new liner almost crept out of the shipyard direct to Southampton. The same strike caused her to leave Southampton an hour before her scheduled time on Thursday 18th August 1960 with Captain G.H. Mayhew CBE in command. She was given a tremendous send off from Southampton and was warmly welcomed on arrival in the Cape ports.

In welcoming the new liner, it was also time to say a fond farewell to "WINCHESTER CASTLE". Since her maiden voyage at the end of 1930, she had served the Company well and also her country with her War time adventures with the Combined Forces. Her First Class Lounge in particular had been very striking with dark wooden panelling, heavy drapes and a marble fireplace with had proved very restful and relaxing. She received spectacular farewells in South African ports, especially at Port Elizabeth where she was saluted by her successor, "WINDSOR CASTLE" as well as "RICHMOND CASTLE", "RIEBEECK CASTLE" and other ships.

Sadly her departure from Southampton was a little of an anti-climax. Normally, our ships depart on their final voyage from the Western Docks and they are saluted by all the ships throughout the port. The traditional salutes for both maiden and final voyages are for the out-going liner to give three long blasts on her siren as she leaves the quay and each ship she passes also give three long blasts which are repeated by the departing ship. As can be imagined, when Southampton Docks is full of shipping, this can be a most impressive send-off. The tones of the sirens can range from the mighty blasts of "QUEEN ELIZABETH" and "UNITED STATES" down to small ships and escorting tugs and are often accompanied by car horns from people watching from the shore.

Unfortunately, in the case of "WINCHESTER CASTLE", she had been berthed at 35 Berth at the far end of the Eastern Docks and out of sight of other shipping. When she eventually sailed she was seen off by shore and office staff, a few tugs and a passing coastal tanker. However, the previous week when the "EDINBURGH CASTLE" sailed, her Master - Captain G.W.B. Lloyd - who had commanded the "WINCHESTER CASTLE" for many years, gave her three hearty salutes.

CHAPTER NINETEEN

NEW CONCEPTS IN SHIPPING

The third new mail vessel to be ordered was the 35,000 ton "TRANSVAAL CASTLE", which was built on the Clyde by John Brown & Company Limited, the biggest vessel to be built there since the "QUEEN ELIZABETH". The keel was laid down on 16th January 1959 and the ship was launched by Lady Cayzer, wife of the Chairman, on 17th January 1961. The liner was fitted with a bulbous bow to economise in power and, of course, fitted with Denny Brown stabilisers. Other facilities were a garage for passengers cars, a laundry for the use of passengers, a medical centre which also included a physiotherapy room and a separate air-conditioned cinema.

"TRANSVAAL CASTLE" was an innovation not only for the Union-Castle Line in particular, but for shipping in general and a new word was coined for her accommodation - Hotel Class. Although the three Intermediate ships ("KENYA CASTLE", "RHODESIA CASTLE" and "BRAEMAR CASTLE") were one-class ships their accommodation and public rooms were standard throughout. The "TRANSVAAL CASTLE" improved on this.

There were no class barriers and the accommodation ranged from single berth cabins to four berth cabins as well as Suites and Cabins-de-luxe. Passengers paid for the type of cabin they required or could afford but, as in a Hotel, once they left the cabins everyone had the full use of all the ship's amenities.

In a yearly magazine published by the Company and available to all passengers called Safari, Mr. Sidney Smith - Group PRO, pointed out all the work involved in creating such a liner. Apart from the actual construction of the ship by John Browns, many other firms throughout the country were also taking part. For example, a firm in Deptford, Kent, were manufacturing the refrigeration machinery, from Glasgow came Echo-sounding machines and life boats, and Whitley Bay provided the mechanical hatch covers. Also in Kent, a Sandwich firm made the generators, from Liverpool came the kitchen equipment and the upholstery was being woven in Dundee. Propellers were manufactured in Birkenhead, electric cable at Leigh in Lancashire, carpets from Durham and furniture at London and Bridgend. The coordination of production of all these items were supervised, as usual, by Mr. Bernard Cayzer and Miss Jean Munro.

Murals were again to play a significant part in the decor. Felix Kelly, who had painted the marvellous pictures of Windsor Castle, was again commissioned to paint murals in the Dining Saloon of the new ship. The theme was a typical South African homestead. Unfortunately, an engine room casing ran up through the Saloon so when passengers entered the Saloon they saw in front of them the entrance to the homestead. Behind the casing was another picture, this time of the rear of the house whilst the tables and chairs in the Saloon were laid out as the garden. Sidney Smith painted a mural in the Assembly Room capturing the Spirit of Venice.

Due to the overall design of the ship, there were other problems involved in the decoration. For example, the Golden Room was over a cargo hatch and in port the centre was taken up by the hatch cover, but this was well disguised. In the Orangery a bar was elegantly designed to form part of the general decor of the room and at the end of each session all stock had to be removed from the bar since shutters would have spoilt the design.

Another innovation was that the Dining Saloon was staffed by waitresses and a new name given to them of stewardettes and a special Hardy Amies uniform was designed for them. This, of course, posed another problem for the planners, the question of separate accommodation. All the ship's Leading Hands had single berth cabins.

The public rooms were luxurious and a complete change from the traditional, even the names, and were designed, again as in a hotel, to meet the varying needs and age groups of the passengers.

Forward on the Promenade Deck was the Smoke Room (the only room to keep its traditional name) and in the centre was a piece of modern art which was designed as a talking point since everyone had their own ideas as to what it represented. Leading aft on one side were the writing room and library whilst on the other side was the Orangery, a light and restful room with its own bar. This led to the

TRANSVAAL CASTLE

largest room in the ship, the Assembly Room, which was the focal point of the ship's entertainment.

Aft of the Assembly Room were the gymnasium and massage room and the Golden Room - a wonderfully designed Verandah Bar overlooking the swimming pool. Its main colour was gold and had a somewhat Eastern design. In the stern of the ship on the Promenade Deck was the Children's Playroom which, at night, doubled as a disco for the younger-at-heart passengers.

On the deck above 'O' (Observation) Deck were the two Suites (named Castor and Pollux) and the cabins-de-luxe. The cabins on this deck were the most expensive on board. There was also a Drawing Room which was used mainly by the older passengers, a Vineyard Bar - a light pleasant room facing the stern and which proved very popular for lunchtime drinks and the Cellar Bar. This was a small intimate bar where one wall contained wine bottle ends cemented into the wall as though in a Cellar. Curtains were pulled each evening forming a very cosy cocktail bar. The bar itself was a double-ended bar serving the Cellar and Vineyard rooms with a two-way mirror to enable the barman to see if anyone was waiting to be served.

There was no dry dock on the Clyde large enough to accommodate the liner for a final service prior to the hand-over ceremony so the "TRANSVAAL CASTLE" sailed for Falmouth for dry-docking before returning to a very foggy Clyde. However, as a launch approached bringing out further crew and stores, the fog lifted and the sun illuminated the ship giving many a first view of her in Union-Castle colours and showing her striking lines.

Shortly before the handing-over ceremony the Managing Director of John Browns was on the radio-telephone link to his office demanding a new hand-over certificate be sent out by launch immediately. Apparently, Sir Nicholas Cayzer's name had been spelt with a 'K'.

On the evening of 17th December 1961 she embarked a party of travel agents, shippers and V.I.P.'s for the run to Southampton and, as they came out in a ferry from Gouroch, a lone piper stood on the top deck of the ferry to pipe them aboard with the sounds of the pipes

DURBAN CASTLE

skirling over the Clyde. She sailed that evening for Southampton passing a very snow-bound Land's End en route.

Whilst at Southampton further groups of agents, shippers etc were entertained on board and, for the first time, a group representing many of the firms who had contributed to the building and fitting-out of the ship.

The "TRANSVAAL CASTLE" sailed on her maiden voyage on 18th January 1962 with Captain A.G.V. Patey in command to the accompaniment of salutes from the various ships in port including the "UNITED STATES". She received a tremendous welcome in the Cape and at Durban over 2,000 local people were shown round the ship.

The "TRANSVAAL CASTLE" replaced the thirty year old "CARNARVON CASTLE" which was withdrawn from service on the 1st June 1962 - she had been the first motor ship in the Fleet and her War-time heroics have already been mentioned. Also withdrawn the same day was the Intermediate ship "WARWICK CASTLE". Built in 1939 and originally called "PRETORIA CASTLE" she was bought by the Admiralty during the War and after buying her back and refitting her for peace-time service she was renamed "WARWICK CASTLE".

As it happened, "CARNARVON CASTLE" and "WARWICK CASTLE" both sailed from Durban together for the last time and the 'two old ladies of the sea' were given a great send off from Durban and other South African ports. In April of that year, the 17,832 ton "DURBAN CASTLE" had also been withdrawn and sold for scrap to a German shipyard - the first British vessel to be scrapped there in post-war years. Apart from the many thousands of miles she had sailed to South Africa and back, in both Mail and Intermediate services, she also had a fine War record including transporting troops to the Sicily, Salermo and Anzio landings and the 'D' Day invasion of the French Riviera.

In 1963 orders were placed for two new Cargo/Mail steamers from Swan, Hunter & Wigham Richardson on the Tyne to be delivered in 1965. Both vessels were to be of 12,000 tons and fitted with twin diesel engines to give a service speed of 22½ knots. The first, "SOUTHAMPTON CASTLE", was launched on 20th October 1964

GOOD HOPE CASTLE

by H.R.H. Princess Alexandra, blessed by the Rt. Rev. K. Lamplugh, Suffragan Bishop of Southampton and was only the second Castle to be named in the Southampton area, the first being "CARISBROOK CASTLE" from the Isle of Wight. Her Commander was Captain A.A. Freer. The original castle was a medieval castle built just inside the City's old walls, but has long since been demolished in favour of a multi-storey block of flats.

The second ship was named "GOOD HOPE CASTLE" and was launched on 16th February 1965 by Mrs. Carol de Wet, wife of the South African Ambassador in London. She sailed on her maiden voyage on 14th January 1966 with Captain D. L'Estrange RD, RNR in command.

Both ships had extensive automation and remote controls in the Engine Room. During their voyage to the Cape they called at the islands of St. Helena and Ascension (under an arrangement with the Commonwealth Office) for which they could accommodate twelve passengers and sometimes deck passengers were carried between the islands. Every member of the crew had a single berth cabin, with a crew mess room run on the style of a cafeteria and there was also a hobby room.

With the entry of these two ships in the Mail Service a new accelerated service commenced with the sailing of "WINDSOR CASTLE" on 16th July 1965 arriving in Cape Town in eleven and a half days at a speed of 22½ knots, the service being maintained by the seven ships.

In the 120 years since the "DANE" first commenced the Mail Service to the Cape, the ships had increased from 530 tons to 38,000 tons and the journey time decreased from forty-four days to eleven and a half days.

CHAPTER TWENTY

CRUISING

Throughout the years Union-Castle has been mainly concerned with regular voyages to South and East Africa. There have been times however, usually to fill in periods between rearranged schedules when the odd cruise had been arranged.

The idea of cruising first started in 1892 when Mr. W.S. Nicholson, a Director of the Union Line, proposed that the Company should institute 'yachting' trips to Hamburg with the Intermediate service. This commenced in 1894 with round voyages from Southampton to Hamburg, Antwerp and Rotterdam. These proved very popular and were continued by the Union-Castle Line until 1914.

When the "KENILWORTH CASTLE" entered the Mail Service in 1904, the "DUNVEGAN CASTLE" was taken off and, before being placed in reserve at Netley, made several cruises to Norway, the Mediterranean and Around Britain. The "DUNNOTTAR CASTLE" went cruising from 1901-1912, but under a charter from Sir Henry Lunn's travel agency.

In the mid 1950's there were a few weeks to spare between the completion of the "KENYA CASTLE" and her advertised sailing so she came to Southampton direct from Belfast and made a cruise to the Atlantic islands. The "WARWICK CASTLE" also undertook several cruises to Spanish ports. Until then this had been the extent of Union-Castle cruising.

A new era dawned in 1963 when an organisation was formed called the Travel Savings Association which was owned by Travel Savings Limited, the principal shareholders being British & Commonwealth, Canadian Pacific (whose service to Canada was in the doldrums) and the Max Wilson Organisation. Its aim was to encourage overseas travel by bringing it within reach of a much wider public and by providing a worldwide range of cruises at moderate prices. It was the reverse of hire-purchase since with T.S.A., future passengers paid so much at a time in advance, and when their payments were

REINA DEL MAR

completed, albeit with great discounts, they could then book their cruise. Four ships were chartered - "EMPRESS OF BRITAIN", "EMPRESS OF ENGLAND", "STRATHEDEN" from Canadian Pacific and the "REINA DEL MAR" from the Pacific Steam Navigation Company Limited, a subsidiary of the Royal Mail Line. All cruises were maintained to the full standard of each ship. In early 1964 the "EMPRESS OF BRITAIN" sailed on an Indian Ocean cruise when Union-Castle Line acted as Agents. At the same time the "REINA DEL MAR", with Union-Castle Line acting as Managers, went to Belfast to make her ideally suitable for cruising including complete air-conditioning and stabilisers.

Her first voyage, later that year under the T.S.A. banner, was from Liverpool to New York for a series of cruises to coincide with the New York Trade Fair. Afterwards she sailed for a cruise to the 1964 Olympic Games in Tokyo via south Africa returning to Southampton (then to be her home port) in May 1965. However, by that time the Max Wilson Organisation had pulled out and the scheme collapsed. The Union-Castle Line took over the charter of what was proving to be a very successful cruise liner, she was painted in Union-Castle colours and embarked on a series of cruises to the North Cape and Mediterranean ports. The "EMPRESS OF ENGLAND" was later sold to Shaw Savill and renamed "OCEAN MONARCH".

The arrival of the "REINA DEL MAR" at Southampton caused a drastic re-think in ship handling. Her projected programme was to arrive early on a Friday morning and sail the same evening on a two week cruise. On the following cruise, after arriving on the Friday morning, she would sail at midday on the Saturday. Passenger Department at Southampton had to deal with these extra passengers as well as the normal 1.00pm sailing of the mail ship. On one Friday alone they handled 1,140 passengers in three hours.

The Catering Department also had its problems with the ordering and shipping of stores. With the mail ship arriving on a Monday, the Chief Buyer (Mr. Jack Elvey) and his staff had several days to calculate and place orders to suppliers before stores were started to be shipped at the end of the week. Even then some stores arrived just before sailing the following Friday. There was not this time for storing "REINA DEL MAR".

The result was that the Chief Catering Officer on board, in consultation with his staff, telexed from the Agents at the last port of call or telegraphed from the ship giving details of stores required for the following cruise, which were then duly ordered. However, there was also the problem of the physical time in actually shipping them. It was therefore decided that all 'wet' stores be delivered to the main supplier, C.G. Hibbert & Company Limited, who delivered them on pallets into the quayside shed ready for shipment. All 'dry' goods were also delivered to the shed, usually the day before arrival.

Regarding the high duty stores (spirits, wines, etc) negotiations were started with H.M. Customs and Excise resulting in a new arrangement. The day before arrival a member of the Wine Section of the Catering Department, a representative of Hibberts and a Customs Officer checked the goods into special metal containers. When they were satisfied that the stock was complete, the Company representative put on padlocks and the Customs Officer official seals. The next morning they were delivered to the quayside and shipped in a matter of minutes. The seals were inspected by Customs and, once the containers were on board, the padlock keys were handed to the Chief Catering Officer and, when at sea, the containers were unlocked and unloaded. The empty containers were off-loaded the next voyage. This system was later adopted by other shipping companies and airlines.

Accommodation on the "REINA DEL MAR" ranged from an outer two-berth cabin with bath to an inner four-berth cabin and were spread over five decks for nearly 1,000 passengers. There were two restaurants on 'C' Deck, the Pacific and the Atlantic - the latter having two sittings. On 'B' Deck were Officers cabins, shop, hospital and consulting rooms. The main entrance was on 'A' Deck together with the Purser's Bureau, another shop and the Dolphin Room and Bar. The main public rooms were on the Promenade Deck above - Mermaid Lounge, swimming pool, Trident Room, Library, Club Room, Ocean Bar and the Coral Lounge - the latter was one of the largest public rooms on a British ship with the longest bar. The Boat Deck contained a sports deck, children's playroom, another swimming pool and Lido with its own bar. On the Navigation Deck above was the ship's cinema. The Ocean Bar, in particular, was very popular since it stayed open until 2.00am and it was known for some passengers not to go ashore sightseeing, but preferred to stay in the Ocean Bar where, they said, they knew what the drinks were like.

Whereas with the usual Mail and Intermediate Services, the Company had been trading at the various ports of call for so long that everything was known regarding facilities etc. However, when planning cruises for the "REINA DEL MAR" at new ports in different countries much information had to be obtained regarding berthing or anchoring facilities, passenger handling, tours, fresh food available etc. She became a very firm favourite with the passengers, many of them returning year after year. She made 12-14 cruises from Southampton each year before sailing to South Africa for two cruises to South America. At first this was followed by another one to Mauritius, but it appeared there was so much competition that, in 1973, the "REINA DEL MAR" came straight back to Southampton after the South American cruises and undertook one to the Caribbean which was so successful it was repeated the following year. Also in 1973, after nine years on charter, the liner was purchased outright by Union-Castle.

However, the cost of maintaining a cruise liner was mounting, especially the cost of fuel and, in 1974, the Company reluctantly announced that they would be withdrawing her on her return from South Africa on the 1st April 1976. Many passengers pleaded with the Directors to change their minds, but to no avail. She eventually sailed from Southampton on 28th May 1976 for breaking up in Taiwan.

In an attempt to bring a more cruising atmosphere, a Cruise Purser (Tourist) was introduced on "WINDSOR CASTLE". The new position was taken by Robbie Rutt, Cruise Purser of the "REINA DEL MAR" who was assisted by Geoffrey Shallcross, 2nd Purser and Peter Glenn, dancing instructor and general entertainer on the cruise ship.

RHODESIA CASTLE

CHAPTER TWENTY-ONE

THE BEGINNING OF THE END

In 1961 the "KENYA CASTLE", "BRAEMAR CASTLE" and "RHODESIA CASTLE" were withdrawn in turn and sent to Belfast for refits, when 12ft was added to their height. They were fitted with 'pepperbox' funnels and their 'A' Deck completely reconstructed with each cabin being provided with a bath or shower. On their return to the Intermediate and Round Africa services, the "DURBAN CASTLE" and "WARWICK CASTLE" were withdrawn and sold for breaking up.

British & Commonwealth ordered three more refrigerated cargo ships in December 1963 from the Greenoch Dockyard Company Limited and built to meet the express requirements of Safmarine who operated them on charter in the South African fruit trade. They were named "S.A. LETABA", "DRAKENSBERG CASTLE" and "S.A. TZANEEN". For the four months of the year when they were not needed as fruit carriers they operated as general cargo steamers.

A few years later, in 1965, the "PRETORIA CASTLE" and "EDINBURGH CASTLE" returned to Belfast where their passenger accommodation was upgraded. They also had their foremasts shortened, the main masts removed altogether and new small masts fitted abaft the Bridge, giving them a more streamlined appearance.

On their return to Southampton the "ATHLONE CASTLE" and "STIRLING CASTLE" were withdrawn following the commencement of the accelerated Mail Service. The two graceful looking 'yachts' of 25,567 tons were built in 1936 and served on the mail run continuously except for War years. They were both sold in 1966 for breaking up, the "ATHLONE CASTLE" in Taiwan and the "STIRLING CASTLE" in Japan. The ship's bell of the "ATHLONE CASTLE" was presented to the Hampshire Cricket Club in Southampton.

In the early part of 1966 there was a National British Seamen's strike which lasted forty-six days, during which time all mail ships were laid up in Southampton together with many other Group vessels

S.A. VAAL

with British crews. It had a devastating effect on British shipping and on the Group's finances.

A result of the strike for shipping enthusiasts, of course, was the sight of so many ships in port at the same time, especially in Southampton where they ranged from "ROXBURGH CASTLE" to "QUEEN ELIZABETH". The local boatmen, with their usual trips round the harbour, had a field day. Some berths had to be kept available for visiting foreign ships and there were cases of ships being double-berthed. There were thirteen of the Group's vessels in port and at 101 Berth in the Western Docks "GOOD HOPE CASTLE", "REINA DEL MAR" and "EDINBURGH CASTLE" were berthed alongside each other. The "ROXBURGH CASTLE" and "ROWALLAN CASTLE" were astern of the "QUEEN MARY" and were even graced with a covered-in gangway previously unknown for a cargo ship.

Throughout the years the progress of the Union-Castle Line has been closely linked with that of South and East Africa, although developments in East Africa had resulted in a reduction of trade for the Company. The connection with South Africa however, had grown stronger and culminated with a decision to run a joint Mail Service with the South African Marine Service (Safmarine). In 1966, the "TRANSVAAL CASTLE" and "PRETORIA CASTLE" were transferred to South African ownership (whilst still manned and managed by Union-Castle).

At separate ceremonies held in Cape Town, later in the year, they were renamed "S.A. VAAL" and "S.A. ORANJE" respectively and sailed in their new colours of white hull, grey funnels with orange, white and blue bands and flying the South African flag.

There was one occasion when this flag had an adverse affect. With a stevedoring strike in Southampton, once passengers had disembarked, the "PENDENNIS CASTLE" sailed for Antwerp to discharge her cargo and to ship stores. Unfortunately, whilst proceeding through the many channels into the port she hit the side of a lock and her return to the UK was delayed. The following week the "S.A. ORANJE" followed and berthed astern of the "PENDENNIS CASTLE" and the crew were assured by Senior Officers of that ship

S.A. ORANJE

from the quayside of the good service provided by the Ports Authorities. Nevertheless, it took nearly two hours before a gangway was in position. It eventually materialised that the Belgians were not happy at the sight of a South African Flag, but after assurances from the crews that the "S.A. ORANJE" was manned by a British crew, the discharge of cargo and shipment of stores ran smoothly.

The "CAPETOWN CASTLE" was later withdrawn from the Mail Service as such, but continued a separate journey from Southampton to Cape Town only and return, calling at the islands of St. Helena and Ascension. She finally left Southampton in 1967 for scrapping in Italy after thirty years service. A plaque in the First Class Long Gallery commemorated her War time service when she sailed 404,000 miles carrying 164,000 troops.

With the commencement of the fruit season that year, the 7,818 ton "ROXBURGH CASTLE" brought the port of Sheerness, in Kent, her first consignment of South African fruit. She was also the largest ship to dock at that port.

The Round Africa and Intermediate services were being affected by high costs and a reduction in trade so, after a service lasting many years, it was discontinued. The "BRAEMAR CASTLE" was delivered to ship-breakers in Faslane, on the Clyde, in January 1966.

The "RHODESIA CASTLE", which it was claimed had travelled 1,080,469 nautical miles carrying 25,000 passengers and thirteen stowaways, left London in 1967 for the Far East. However, she helped make revenue to the end. She left the River Blackwater on 13th July 1967 for Rotterdam where she loaded a bulk cargo for India ports via the Cape. She discharged her cargo in Bombay where she also bunkered before proceeding to Formosa for breaking up in the November. She was followed the following month by the oldest and smallest ship in the Fleet, "ROSLIN CASTLE". The "KINPURNIE CASTLE" (ex "S.A. SPRINGBOK") was sold for further trading as the "HELLENIC MED".

In March 1967 the "KENYA CASTLE" (which was built in 1952) was also sold for further trading to the Greek Chandris Line who renamed her "AMERICANIS" and in early 1992 was still trading with them, under the title of Fantasy Cruises, in the Caribbean giving excellent food and service.

A serious fire broke out in the Engine Room of the "GOOD HOPE CASTLE" on 29th June 1973 whilst en route from Ascension to St. Helena. The original outbreak was brought under control, but later spread to the accommodation. The crew, under Captain R.J. Miller, fought the fire all night but by morning the passengers and crew were sent away in the life boats and an hour later Captain Miller and the rest of the crew followed. They were all picked up by a tanker and landed at Ascension. Technical staff were flown out from the UK to join the crew together with that of the "CLAN MALCOLM" which was standing by to salvage the ship which was then towed to Antwerp for inspection. In the September she was towed to Bilbao for repairs. She re-entered the Mail Service on the 31st May 1974. Two years later she and her sister ship "SOUTHAMPTON CASTLE" were featured in a new issue of stamps by the islands of St. Helena, Ascension and Tristan de Cunha.

British & Commonwealth then announced that the "S.A. ORANJE" and "EDINBURGH CASTLE" would shortly be withdrawn. This would leave the three new passenger liners, the two cargo/mail liners and two fast cargo ships to maintain the service.

When the "S.A. ORANJE" left the Cape for the last time on 3rd October 1975 a special first day stamp cover was issued with a photograph of the ship and one of Field Marshall Smuts, whose wife had launched the ship.

The "EDINBURGH CASTLE" sailed from Southampton to the Cape in March 1976 on her last and 189th voyage and as she left she was piped away from the quayside by two Union-Castle officers - Bill Maitland - Engineer and Bill McIntosh - Radio Officer. She returned to her home port on 12th April and shortly afterwards sailed to the Far East for breaking up. En route she called in to Durban where she donated her No. 4 life boat to the South African Natal Cadet Corps.

CHAPTER TWENTY-TWO

THE END OF THE LINE

By December 1973 world oil prices had rocketed sharply and shipping companies were forced to look at ways of economising in fuel and to increase their revenue. Union-Castle announced that their mail ships would slow down and take an extra day on each leg of the journey from Southampton to Durban. Passenger fares were also increased.

Late in 1974 it was announced that cargo operations between Europe and South Africa would be containerised during 1977-1978 and that the Union-Castle and Clan Line's interests would be handled by Overseas Containers Limited in whom the Group had a financial interest.

Finally, it was announced, with regret, that the Mail Service would be discontinued by the end of September 1977. The "WINDSOR CASTLE" and "S.A. VAAL", in particular, were built to carry several thousand tons of cargo which would now be containerised and the two ships would not be viable carrying passengers only.

The first ship to be affected by high wages and heavy fuel bills was the "PENDENNIS CASTLE" in June 1976, earlier than expected. At the same time it was announced that on her withdrawal, Union-Castle ships would no longer call at Madeira. When the island was first discovered in 1419 it was uninhabited, but was later populated and one or two Union Line and later Union-Castle Line vessels had called in at the island in the early days, but it was not until after the turn of the Century that regular calls were made. With its mild climate and beautiful scenery, Madeira started to attract more English people and in the 1930's its popularity as a winter resort had increased so much that a member of the Passenger Department staff in London was sent there to assist the Agents, Blandy Bros., whose help for the Company has been invaluable over the years.

First Class Cabin - WINDSOR CASTLE

The "PENDENNIS CASTLE" sailed from Southampton on 7th May 1976 on her last and 124th voyage to the Cape under the command of Captain J.S. Catterall and when she finally left Cape Town on 2nd June she was given a rousing send off from all the ships in port. On arrival at Southampton, after steaming a total of 1,675,000 nautical miles, she was flying a paying-off pennant for her eighteen years of service. She finally left her home port in the July for a journey to the Far East via the Suez Canal (the first time a Company's ship had used the Canal for many years) and briefly anchored off Aden. She was sold for further trading in the Far East and was last seen at anchor off Hong Kong under the name of "OCEAN QUEEN", but it is believed that this project did not work out and she was later scrapped. To take her place in the Mail Service, the 9,784 ton "ANDALUCIA STAR" was time-chartered from the Blue Star Line.

The "S.A. VAAL" slightly extended her Southbound voyage in May 1966 by embarking the Prime Minister of South Africa - The Hon. B.J. Voster - the Leader of the Opposition and other notables, and, escorted by the frigate "PRESIDENT KRUGER", sailed to Richards Bay, a new port north of Durban, for the official opening ceremony.

The "WINDSOR CASTLE" sailed from Southampton on her last voyage to South Africa on 12th August 1977 and during the voyage, Sir Nicholas Cayzer received a message from H.M. The Queen Mother, who had launched the ship in 1959. She asked him to convey to the Master and the Ship's Company how much they were in her thoughts at the completion of the voyage and sent them her warmest good wishes.

Her departure from Cape Town at exactly 4.00pm on Thursday 6th September was a very emotional scene. Escorted by four tugs, dressed overall, she was saluted by all the ships in port, with a military band playing 'Auld Lang Syne' on the quayside and aircraft from the Maritime Command flying overhead. She was escorted out by the South African Naval Frigate "PRESIDENT PRETORIUS" who, when the liner was out into Table Bay, fired a shot across her bows and ordered her to heave-to. A party of 'pirates' boarded the

"WINDSOR CASTLE", dashed up to the Bridge and presented Captain P.St.Q. Beadon with an embroidered mail bag. She arrived back at Southampton on 19th September 1977 - one hundred and twenty years and four days since the tiny 530 ton "DANE" sailed down Southampton Water on the first mail run to the Cape. "WINDSOR CASTLE" was sold to a firm in the Middle East for use as an hotel ship and renamed "MARGUERITA C".

The "S.A. VAAL" was given a tremendous send off by the people and port of Southampton when she sailed on her last voyage to the Cape, one which was echoed at all her ports of call. At Durban the Publicity Association had arranged a spectacular show on the quayside with the Caledonian Pipers, Durban's Men's Choir and Zulu dancers. At Cape Town thousands of people lined the shore to watch her depart for the last time, also the last Union-Castle passenger vessel to leave the Cape. She arrived back in southampton on 10th October 1977 and was sold to the Carnival Shipping Company and is still successfully cruising in the Caribbean as the "FESTIVAL".

The "GOOD HOPE CASTLE" had arrived back in Southampton on 26th September so it was left to the "SOUTHAMPTON CASTLE" to be the last ever mail ship to South Africa and when the signal 'Finished with Engines' was given on arrival at Southampton on 24th October 1977 it also signalled 'THE END OF THE LINE'.

So the Mail Service which started 120 years earlier due to the foresight and inspiration of Arthur Anderson and continued by Donald Currie and the Cayzer family came to a close due, in the main, to the modern inventions of aircraft and containers. Many ships have come and gone and their names will still live long in memories. It is a fascinating story of strength and determination in the early days and the loyalty of the shore staff and ship's crews throughout the years.

The final comment must be a poem by Mr. K.J. Brett, formerly of the Passenger Department in London, which seems to sum up the emotions felt at the end of the era:

Salute to the Castles

Castles are our memories, Castles are our pride,
Sadly now we mark their passing day,
Remembering the tall ships straining on the tide,
And bringing up the dawn in Table Bay.

Southward gleam the seaways, seas they knew so well,
Buffeting through Biscay, blistered in the sun,
A long, long line of Castles, dipping with the swell,
To keep the weekly bargain on the great Cape Run.

Castles were our calling, Castles were our trade,
Our purpose and our passion and our creed,
They were part of history whose story will not fade,
Helping to build a Continent, serving a Nation's needs.

The Castles brought the lifeblood, the people, the mails,
The rich, the poor, the restless, and the great,
Cotton goods and Paris hats, machines and iron nails,
Eternal as the seasons and hardly ever late.

From Isle of Wight and Durban bluff, Sea Point and Signal Hill,
Year in, year out they greeted the Castles steaming by,
Each one a dear familiar friend and all remembered still,
For ships and men may come and go but honour does not die.

From Windsor to Balmoral, Athlone to Armadale,
Cape Town, Kinfauns, Llanstephan, the names ring out like bells,
A hundred years they sailored and did not fault or fail,
Here's a toast to all proud ships as we say our sad farewells.

CHAPTER TWENTY-THREE

EPILOGUE

In 1977 in an endeavour to keep alive the names of some of the old Union-Castle ships, four Clan Line ships, "ROSS", "RAMSEY", "RANALD" and "ROBERTSON" were renamed into Castles, "KINPURNIE", "WINCHESTER", "DOVER" and "BALMORAL". These were later transferred to the Universal Reefer Company and were later still joined by "EDINBURGH UNIVERSAL" and "STIRLING UNIVERSAL".

Whilst the Union-Castle fleet had decreased, that of the King Line had increased with bulk carriers and tankers and with the remaining ships of the Group were managed by Cayzer Irvine Shipping Company Limited under the direction of Mr. J. Andrea and Captain Reg. Kelso (one of the most popular Masters of the "REINA DEL MAR"). However, in December 1981 the last Clan Line ships were sold and by 1986 British & Commonwealth had disposed of their last ship and was now renamed B. & C. Holdings plc.

Sir Nicholas Cayzer received a Life Peerage in the New Year's Honours List in 1982 taking the title of Lord Cayzer of St. Mary Axe in the City of London. However, this was tinged with sorrow at the death of his brother, Mr. Bernard Cayzer, at his home in Bath five days earlier at the age of sixty-seven. Mr. Bernard Cayzer who had long experience of shipping in Liverpool and Glasgow had been responsible for passenger and catering services throughout the Group. It was under his guidance that the almost revolutionary standards of decor and accommodation of the last three Passenger Mail ships received International acclaim in the shipping world.

Sir George Christopher, former Chairman of Union-Castle Line, died on the 24th November 1978 aged eighty-seven.

For many years, British & Commonwealth had been diversifying into other industries, financial services, airlines, helicopters, office furniture, etc. In 1987 the commercial and service companies were amalgamated into a new Company, Bricom plc, whilst the remaining companies, mainly in the financial sector, continued with British & Commonwealth.

Lord Cayzer, Lord Rotherwick and the Hon. Anthony Cayzer all retired from the Board of British & Commonwealth and Mr. John Gunn took over as Chairman. Sadly, shortly afterwards, the Company collapsed and the trading of their shares was discontinued on the Stock Exchange.

INDEX OF UNION-CASTLE SHIPS